Cities in a Sunburnt Country

As Australian cities face uncertain water futures, what insights can the history of Aboriginal and settler relationships with water yield? Residents have come to expect reliable, safe, and cheap water, but natural limits and the costs of maintaining and expanding water networks are at odds with forms and cultures of urban water use. *Cities in a Sunburnt Country* is the first comparative study of the provision, use, and social impact of water and water infrastructure in Australia's five largest cities. Drawing on environmental, urban, and economic history, this co-authored book challenges widely held assumptions, both in Australia and around the world, about water management, consumption, and sustainability. From the 'living water' of Aboriginal cultures to the rise of networked water infrastructure, the book invites us to take a long view of how water has shaped our cities and how urban water systems and cultures might weather a warming world.

Margaret Cook is an history lecturer at the University of the Sunshine Coast, author of *A River with a City Problem: A History of Brisbane Floods* (2019) and co-editor (with Scott McKinnon) of *Disasters in Australia and New Zealand* (2020). She was the recipient of the John and Ruth Kerr Medal of Distinction in History (2020).

Lionel Frost is the author of *The New Urban Frontier: Urbanisation and City-Building in Australasia and the American West* (1991), winner of the Dyos Prize in Urban History (1994), and a contributor to the *Cambridge History of Australia* (2013), *Cambridge World History* (2015), and *Cambridge Economic History of Australia* (2015).

Andrea Gaynor is Professor of History and Australian Research Council Future Fellow at The University of Western Australia. An environmental historian, she seeks to use the contextualising and narrative power of history to support transitions to more just and sustainable societies.

Jenny Gregory, AM FRHS, is Emeritus Professor of History at The University of Western Australia. Her research focuses on urban history, primarily town planning and heritage. Her books include *City of Light: A History of Perth since the 1950s* (2003) and, as editor-in-chief, the *Historical Encyclopedia of Western Australia* (2009).

Ruth A. Morgan is an environmental historian at The Australian National University, whose prize-winning work on the histories of water and climate has been generously funded by the Australian Research Council and the Alexander von Humboldt Foundation. She is a lead author in Working Group II of the Intergovernmental Panel on Climate Change's Sixth Assessment Report.

Martin Shanahan is Professor of Economic and Business History at The University of South Australia and Elof Hansson Visiting Professor in International Business and Trade at Gothenburg University, Sweden. A recipient of the Butlin Prize in Economic History, he has also written on wealth and income distribution, international cartels, and water markets.

Peter Spearritt is Emeritus Professor in History at The University of Queensland and Fellow of the Academy of Social Sciences in Australia. His books include *Sydney's Century* (2000), *Where History Happened: The Hidden Past of Australia's Towns and Places* (2018), and, as co-author, *The Twentieth-Century Historic Thematic Framework* (2021).

Studies in Environment and History

Other Books in the Series

Adam Sundberg, *Natural Disaster at the Closing of the Dutch Golden Age: Floods, Worms, and Cattle Plague*
Germán Vergara, *Fueling Mexico: Energy and Environment, 1850–1950*
David Moon, *The American Steppes: The Unexpected Russian Roots of Great Plains Agriculture, 1870s–1930s*
James L. A. Webb, Jr., *The Guts of the Matter: A Global Environmental History of Human Waste and Infectious Intestinal Disease*
Maya K. Peterson, *Pipe Dreams: Water and Empire in Central Asia's Aral Sea Basin*
Thomas M. Wickman, *Snowshoe Country: An Environmental and Cultural History of Winter in the Early American Northeast*
Debjani Bhattacharyya, *Empire and Ecology in the Bengal Delta: The Making of Calcutta*
Chris Courtney, *The Nature of Disaster in China: The 1931 Yangzi River Flood*
Dagomar Degroot, *The Frigid Golden Age: Climate Change, the Little Ice Age, and the Dutch Republic, 1560–1720*
Edmund Russell, *Greyhound Nation: A Coevolutionary History of England, 1200–1900*
Timothy J. LeCain, *The Matter of History: How Things Create the Past*
Ling Zhang, *The River, the Plain, and the State: An Environmental Drama in Northern Song China, 1048–1128*
Abraham H. Gibson, *Feral Animals in the American South: An Evolutionary History*
Andy Bruno, *The Nature of Soviet Power: An Arctic Environmental History*
David A. Bello, *Across Forest, Steppe, and Mountain: Environment, Identity, and Empire in Qing China's Borderlands*
Erik Loomis, *Empire of Timber: Labor Unions and the Pacific Northwest Forests*
Peter Thorsheim, *Waste into Weapons: Recycling in Britain during the Second World War*
Kieko Matteson, *Forests in Revolutionary France: Conservation, Community, and Conflict, 1669–1848*
Micah S. Muscolino, *The Ecology of War in China: Henan Province, the Yellow River, and Beyond, 1938–1950*
George Colpitts, *Pemmican Empire: Food, Trade, and the Last Bison Hunts in the North American Plains, 1780–1882*

(Continued after the Index)

Cities in a Sunburnt Country

Water and the Making of Urban Australia

MARGARET COOK
University of the Sunshine Coast

LIONEL FROST
Monash University

ANDREA GAYNOR
The University of Western Australia

JENNY GREGORY
The University of Western Australia

RUTH A. MORGAN
The Australian National University

MARTIN SHANAHAN
The University of South Australia

PETER SPEARRITT
The University of Queensland

CAMBRIDGE
UNIVERSITY PRESS

CAMBRIDGE
UNIVERSITY PRESS

University Printing House, Cambridge CB2 8BS, United Kingdom

One Liberty Plaza, 20th Floor, New York, NY 10006, USA

477 Williamstown Road, Port Melbourne, VIC 3207, Australia

314–321, 3rd Floor, Plot 3, Splendor Forum, Jasola District Centre, New Delhi – 110025, India

103 Penang Road, #05–06/07, Visioncrest Commercial, Singapore 238467

Cambridge University Press is part of the University of Cambridge.

It furthers the University's mission by disseminating knowledge in the pursuit of education, learning, and research at the highest international levels of excellence.

www.cambridge.org
Information on this title: www.cambridge.org/9781108831581
DOI: 10.1017/9781108917698

First published 2022

A catalogue record for this publication is available from the British Library.

ISBN 978-1-108-83158-1 Hardback

Contents

Figures

Maps

* These maps have been crafted critically by Daniel Jan Martin, drawn together at the same scale and with the same information, to facilitate a rare cartographic comparison between Australian cities, their water, and their landscape structures.

Tables

Acknowledgements

Research for this book was supported by the Australian Research Council's Discovery Projects scheme (DP180100807).

We are grateful to Daniel Jan Martin for creating the maps and for conversations about wetlands, Indigenous water, and numerous other topics of relevance to urban water history. Our sincere thanks to Susan Avey, Nathan Etherington, Elizabeth Gralton, Adam Long, and Dmytro Ostapenko for their invaluable research assistance. Thanks also to Isa Lacuna for preparing the index and to Nick Cook for formatting assistance.

Earlier versions were presented at the International Water History Conference in Delft 2015, World Economic History Congress in Boston 2018, Asia-Pacific Business and Economic History Conference in Canberra 2019, and Australian Historical Association Conference in Toowoomba 2019. We thank participants for their feedback. We are grateful to the Rachel Carson Center for Environment and Society for hosting a virtual exhibition of our work, 'Drought, Mud, Filth, and Flood: Water Crises in Australian Cities, 1880s–2010s' (www.environmentandsociety.org/exhibitions/drought-mud-filth-and-flood), on which this book builds. We also thank the Cooperative Research Centre for Water Sensitive Cities for supporting an earlier project, led by Associate Professor Seamus O'Hanlon, which laid the groundwork for the research on which this book is based.

For advice on sources, feedback, and support, we are grateful to Matthew Barton, Frank Bongiorno, Luc Borrowman, Julie Collins, Lauretta Davies, Graeme Davison, John Doody, Charles Fahey, Jennifer Frost, Warwick Frost, Christine Garnaut, Vyonne Geneve, David

Goldfield, Sue Graham-Taylor, Abdel K. Halabi, Damien Hassan, Lee-Anne Khor, Andrea Kirk-Brown, Seamus O'Hanlon, Philip Light, Annabel Lloyd, Clare Menck, David Merrett, Char Miller, Baden Myers, Fiona Newton, Robert Noffke, Emily O'Gorman, Bobbie Oliver, Gary Osmond, Robert Pascoe, Simone Sharpe, Claire Smith, Phoebe Tang, John Taylor, Robyn Taylor, and John Viska.

For access to information and images, we thank the Brisbane City Council Archives, Museum of Applied Arts and Sciences (Sydney), Queensland State Archives, State Library of Western Australia, State Records of Western Australia, and National Library of Australia.

Thanks to Rachel Bleifeder and Lucy Rhymer, and to Emily Sharp, John R. McNeill, and Ling Zhang, general editors of the Studies in Environment and History series, and the Press Syndicate of Cambridge University Press.

Margaret Cook: I have thoroughly enjoyed collaborating on this project with such skilled, knowledgeable, and generous historians. Thanks to the team for their camaraderie, to Lionel Frost for his patient project management, and to Andrea Gaynor for her unwavering leadership. My love to Nick, Andrew, and Cameron for their constant support and encouragement.

Lionel Frost: Thanks to my co-authors for their wisdom and generosity. I would like to acknowledge my debt to three mentors, now deceased, who guided my early steps as an urban historian: John McCarty, Donald J. Olsen, and Tony Sutcliffe. Thanks to Mark and Jacinta, Aaliyah, Robert, Monica and Adam, Anaka, and Hank for your love, and always having my back. Thanks and love to Maree.

Andrea Gaynor: Sincere thanks to all co-authors and research assistants on this project for their patience, good humour, and insights as we journeyed together in a turbulent time towards a complete manuscript. In particular I would like to thank Lionel Frost for his project coordination and Margaret Cook for her work on the referencing and images, in addition to her formidable intellectual input. Many thanks also to Jamie, Otto, Griff, and Emily for their support and encouragement throughout the book project.

Jenny Gregory: Thanks to a dream team for their insights, collegiality, and friendship. Sincere thanks to Andrea Gaynor and Lionel Frost for their leadership, and to Ross always for his knowledge, encouragement, and tolerance.

Ruth A. Morgan: It was an honour to work with such an incredible team of historians to bring together this manuscript under challenging circumstances. I am especially indebted to Lionel Frost, Margaret Cook, and Andrea Gaynor for their coordination and guidance over the course of the project. With thanks and love to Tash for her patience, support, and encouragement.

Martin Shanahan: I would like to acknowledge my colleagues on this project who have taught me a great deal while also having fun. I would also like to thank the loving support of my family and especially my father, Ross, who set me off on this path of adventure.

Peter Spearritt: Four deceased mentors encouraged me to take cities as a worthy topic of study: John Iremonger, Heather Radi, Tom Stannage, and Patrick Troy. Anne Gilmore accompanied me on field trips to Warragamba Dam, the Nepean floodplain, the Hinze Dam, and the Tugun Desalination Plant.

1

Prologue

I love a sunburnt country,
A land of sweeping plains,
Of ragged mountain ranges,
Of droughts and flooding rains.
Dorothea Mackellar, 'My Country' (1908)[1]

Dorothea Mackellar's poem 'My Country', first published in London in 1908 by its homesick author, recalls life in rural Australia. Born into a wealthy Sydney family in 1885, Mackellar's Australian-born parents owned a number of large rural properties.[2] The poem opens by suggesting that 'the love of field and coppice' and of 'ordered woods and gardens' runs in the veins of a British audience. Mackellar regrets Australia's 'ring-barked forests' but celebrates 'the wide brown land' and declares love for the 'sunburnt country'. 'Droughts and flooding rains' are the essence of the water management challenges across Australia, including its ever-growing cities. When Mackellar wrote 'My Country', 55 per cent of Australians lived in cities and towns, many of which were vulnerable to extremes of too little water, or too much. The five mainland state capitals

[1] Mackellar, *My Country and Other Poems*, 11–12.

[2] Beverley Kingston, 'Mackellar, Isobel Marion Dorothea (1885–1968)', *Australian Dictionary of Biography*, National Centre of Biography, Australian National University, https://adb.anu.edu.au/biography/mackellar-isobel-marion-dorothea-7383/text12835, published first in hardcopy 1986, accessed 16 April 2021; *Sydney Mail* 21 October 1908. The poem first appeared in the *Spectator* under the title 'Core of My Heart', but when published in book form the title became 'My Country'.

alone were home to one in three Australians; currently the figure is two in three.

How has water shaped the largest Australian cities? What are the historical drivers that produced today's urban water systems, and how have these systems impacted on human and ecological welfare? This book aims to reveal not only why our urban water systems developed as they did but also the many ways in which they have shaped city growth and suburban development.

There is much to learn from the stories of how both Indigenous and non-Indigenous Australians have learned to live with climate variability over a range of timescales. Mackellar's poem makes no mention of Indigenous inhabitants or landscapes, but it is important to recognise that the places that are now Australia's five largest metropolitan capitals were deeply known, managed, and loved for tens of thousands of years before the present, and that the Aboriginal peoples of those places have – remarkably, given the violent disruption of colonisation – maintained connection with and knowledge of those places. In telling of Aboriginal relationships with water in these locations, our approach as non-Indigenous historians is to collate narratives respectfully from Indigenous and non-Indigenous accounts published in the public domain, whether in the form of oral testimony or written word. Where possible, we follow the guidance of Palyku legal scholar Ambelin Kwaymullina by seeking to privilege Indigenous voices and engage with works produced by Indigenous people in ethical partnerships with non-Indigenous people.[3] We recognise that while there is a diversity of Indigenous views, 'Aboriginal stories, however expressed or embodied, hold power, spirit and agency'. Moreover, 'Knowledge can never be separated from the diverse Countries that shaped the ancient epistemologies of Aboriginal people, and the many voices of Country speak through the embodiment of story into text, object, symbol or design.'[4] We pay our respects to the First Nations people and the land on which their ancestors lived for millennia. We hope this current work reflects, in some small part, their generations of custodianship and a perspective that will endure and guide future generations.

In 2015, the World Economic Forum declared water crises to be the biggest risk facing the planet.[5] Yet, in Australia, there has been

[3] Kwaymullina, 'Research, ethics and Indigenous peoples'.

[4] Kwaymullina, Kwaymullina and Butterly, 'Living texts', 5.

[5] World Economic Forum, *Global Risks Report 2015*, http://reports.weforum.org/global-risks-2015/, accessed 23 April 2021.

surprisingly little reflection on how the past has shaped our urban water systems, or how this knowledge might inform present solutions to water management problems. There is an extensive literature on the history of Australian urbanisation, including studies of individual cities and histories of particular suburbs or local government areas.[6] The issue of how governments overcame the challenge of providing water to urban residents is implicit in much of this literature. Yet, these works provide no coherent understanding of how water availability and scarcity have shaped urban structures and environments, neighbourhoods' social experiences, or individual living standards through different periods of population growth, economic and land use changes, flood, and drought. The historiography of water in Australia has largely focused on the rural sector and irrigation schemes.[7] Many book-length studies of urban water have been commissioned by the statutory authorities that built the dams, reservoirs, pipelines, and sewage treatment plants.[8] Very little has been written on how the majority of Australians, the capital city dwellers, have shaped and been shaped by one of the nation's most defining characteristics – its relationship to fresh water.

Cities in a Sunburnt Country is the first comparative study of the provision, use, and social impact of water and water infrastructure in Australia's five largest cities – the nation's only cities to have exceeded a population of 1 million. Large-scale water supply and sewerage infrastructure has underpinned the growth of large cities, with intended and unintended consequences. Beginning with the First Australians, the book sketches Indigenous understandings and approaches to water and Country, before turning to the early colonial development of small-scale water sources and then the domestication of water with large-scale water supply and sewerage infrastructure. The book concludes in the twenty-first century, when prolonged drought and the spectre of anthropogenic climate change raised questions as to the sustainability of long-standing assumptions and strategies for urban water management.

6 See for example Frost and O'Hanlon, 'Urban history and the future of Australian cities'; O'Hanlon, *City Life*; Davison, *City Dreamers*.

7 See for example Cathcart, *The Water Dreamers*; O'Gorman, *Flood Country*; Anderson, *Endurance*; Jones, *Slow Catastrophes*.

8 See for example Aird, *The Water Supply*; Dingle and Rasmussen, *Vital Connections*; Powell, *Watering the Western Third*; Lloyd, *Either Drought or Plenty*. Troy (ed.) *Troubled Waters* brings together historians, geographers, ecologists, and lawyers to examine the Millennium Drought-induced urban water crisis.

As Australia's cities continue to grow in the face of changing climates, water has been thrust to the forefront of national conversations about sustainability and urban futures. Fresh water resources within easy reach have already been harnessed, often to their maximum capacity, attaching a mounting sense of urgency to these discussions. An understanding of why complex water cultures and infrastructures exist in their current forms can enable a more historically informed and equitable approach to urban water provision and management. While new approaches and technologies will be required to face future challenges, and there is something deeply attractive about models of well-designed water-sensitive cities, plans to dramatically alter urban water systems must not only work with the present urban fabric but also rely on public support. Examining the development of urban water supply can help citizens to reflect upon what they value about present systems and inspire confidence to change without losing what counts most. Policymakers, too, can consider how past changes have occurred, in order to inform their own approaches to developing just, equitable, and sustainable strategies for managing and accommodating water in Australian cities.

At a time when water systems of large cities worldwide are under pressure from environmental change and population growth, historical knowledge of the provision, use, and cultures of water is critical. Comparatively rapid changes to water supplies, river flows, the frequency of floods and droughts, and rising sea levels mean the need for cities to respond effectively to the challenge of water supply and sanitation is stronger than ever. Every metropolis produces its own set of water management problems that must be addressed in the context of particular knowledges, cultures, technologies, and levels of economic development for the city to continue to function. Some decisions, once made, cannot easily be undone – and may have ongoing consequences for later generations. The challenge to produce effective water management strategies that foster more sustainable, resilient, productive, liveable, and equitable cities is an example of a 'wicked' problem that defies simple solution.

This book was written by seven researchers with a shared mission of seeking to contribute to the creation of more resilient and sustainable water systems through an enhanced understanding of water history. Its genesis lies in an invitation to join the Cooperative Research Centre (CRC) for Water Sensitive Cities, formed in 2012 to bring together researchers from disciplines and sub-disciplines across social and biophysical sciences and humanities. This interdisciplinary forum allowed us to share ideas and approaches, with dialogue formalised through works of synthesis, including a summary CRC report,[9] as well as a special issue of *Journal of Urban History* ('Water and the making of Californian and Australian cities').[10] In 2015, we secured funding from the Australian Research Council to produce the first longitudinal, multi-city comparative historical study of how and why water has been provided to Australian cities, and how it has transformed them. One of the principal aims of the new project was to integrate cultural, economic, and environmental histories. We felt that this was an essential precondition to two of the other project aims: producing new knowledge that might help to improve the effectiveness, equity, and sustainability of Australian urban water policy; and producing water stories that engage the public through history and help us imagine alternatives. In addition to this book, the project also produced a virtual exhibition at the Rachel Carson Center[11] and a special issue of *Australian Historical Studies* ('Ripple effects: Urban water in Australian History').[12]

As a 'wicked' problem, the achievement of sustainable water systems requires bridging disciplinary divides to deliver knowledge that can inform appropriate resource use and policy action. This necessity is increasingly recognised in the broadening field of urban water management, which has extended beyond engineers and sanitarians to include economists and more recently the insights of the humanities and social sciences, as well as Indigenous peoples. Our team also brings together diverse expertise and perspectives. As individual scholars, we have published in environmental history, economic history, urban history, planning history, and public administration. Some of us have published in more than one of those fields. As historians from different sub-disciplines working in conversation across our various areas of expertise, we are able to ask new questions, identify the complex causes of past problems,

9 Frost, Gaynor, Gregory, Morgan, O'Hanlon, Spearritt, and Young, *Water, History and the Australian City: Urbanism, Suburbanism and Watering a Dry Continent, 1788–2015* (Melbourne: Cooperative Research Centre for Water Sensitive Cities, 2015), https://watersensitivecities.org.au/wp-content/uploads/2016/12/A2.1_2_2016_R5-19-12-2016-V2.pdf, accessed 20 June 2021.

10 See Frost, 'Water technology and the urban environment'; Dobbie, Morgan and Frost, 'Overcoming abundance'; Gaynor, 'Lawnscaping Perth'; Gregory, 'A spirit of Bolshevism'; O'Hanlon and Spearritt, 'From water engineers to financial engineering'.

11 Drought, Mud, Filth, and Flood: Water Crises in Australian Cities, 1880s–2010s (Rachel Carson Center, Virtual exhibitions 2019, no. 3). www.environmentandsociety.org/exhibitions/drought-mud-filth-and-flood, accessed 1 June 2021.

12 O'Gorman and Morgan, 'Fluid terrains'; Frost and Shanahan, 'Domesticating water; Gaynor and Martin, 'Every fountain tells a story'; Cook and Spearritt, 'Water forever'.

unearth a wide range of relevant stories, and suggest diverse lessons for the future. Accordingly, rather than an anthology of chapters from individual authors – reflecting author sub-disciplinary perspectives – each chapter has been co-written. Each is the product of constructive dialogue and co-creation between researchers who have developed and drawn on both disciplinary expertise and interdisciplinary orientation.

Australia is the driest inhabited continent, with a desert centre and few large river systems. Located on coastal fringes and separated by long distances, the five cities are marked by climatic diversity. Adelaide and Perth's Mediterranean climates are comparable to those of the Mediterranean region itself, and other regions where a similar climate prevails, such as southern California, south-central Chile, and South Africa's Western Cape. Such climates are confined to relatively narrow coastal fringes, as mountain ranges block the movement of oceanic weather fronts, creating rain shadow effects. In nearby mountains and hills where rainfall is higher, river systems and valleys in nearby mountains provide sites for urban water storages to supply cities on the coastal plains.

Perth's long-term average (since 1876) annual rainfall of 838 mm is much higher than that of Adelaide, which averages 533 mm.[13] Perth, however, has experienced a drying trend from the last quarter of the twentieth century. Average rainfall has been declining since 1971, and the figure from 1994 to 2015 was 13 per cent below the historic average. Very little rain falls in Perth in summer, the sandy soil does not retain water, and evaporation rates are high. In Adelaide, the average evaporation rate exceeds rainfall every month except in late autumn and winter. That city's water supply is critically dependant on the River Murray and the River Murray basin – a semi-arid region that has the lowest run-off of any large river system in the world.[14] The flow of water through the river system exhibits a wider range of variability than almost any other of all the world's major river systems.

Melbourne's climate is more temperate and oceanic, with changeable weather conditions resulting from the influence of hot inland winds and air masses and the cool Southern Ocean. Rainfall averages 521 mm per annum, distributed more evenly over the year than in Perth and Adelaide.

[13] Australian Government, Bureau of Meteorology, 'Climate statistics for Australian sites www.bom.gov.au/climate/averages/tables/ca_site_file_names.shtml, accessed on 7 June 2021.

[14] River Murray Commission, *Harnessing Australia's Greatest River*, 34–35.

The more northerly east coast capitals, Sydney and Brisbane, have temperate and sub-tropical climates with much higher average rainfall than the three southerly cities (1175 mm per annum in Sydney, 1149 mm in Brisbane), although both are also vulnerable to periodic, extended drought and significant variation in average rainfall across the urban area. Sydney has a temperate climate of warm, sometimes hot summers and mild winters, with temperature variations moderated by proximity to the Pacific. Brisbane's climate is sub-tropical, with hot, humid summers, in which thunderstorms are common, and dry, moderately warm winters.

The east coast capitals are subject to the effects of the El Niño-Southern Oscillation (ENSO), alternating periods of seven to ten years of dry and wetter conditions. At longer, unpredictable intervals of twenty to thirty years, they may also be impacted by Pacific Decadal Oscillation patterns of warming and cooling of the ocean surface, which act as 'throttle and brake' for shorter ENSO cycles of drought and flooding rains.[15] Perth's rainfall is also affected by variability in sea surface temperatures associated with the Indian Ocean Dipole.

Melbourne, Brisbane, Perth, the north-western suburbs of Sydney, and parts of Adelaide are built on floodplains. Melbourne's Yarra River drains a catchment of over 5000 square kilometres south of the Great Dividing Range, a complex of mountains and uplands extending from western Victoria to northern Queensland that separates coastal areas and inland grasslands. The city of Melbourne is located on flat, naturally swampy ground. Major floods occurred in Melbourne in 1839, 1863, 1891, and 1934. At around 13,500 square kilometres, the Brisbane River catchment, also in the Great Dividing Range, is larger than the Yarra catchment and subject to heavier rainfall. Brisbane's inundation in 1893 was the result of three flood events within one fortnight, the 1974 flood was caused by rainfall from decaying tropical cyclones, and the catastrophic flood of 2011 was associated with strong La Niña rain events.[16] Perth's Swan River and its tributaries are part of a very large river system, the Avon River, which has a catchment area of almost 120,000 square kilometres. Perth's most extensive flooding was in 1862, 1872, and 1926; the city experienced eleven floods with an average recurrence interval of ten years between 1910 and 1983.[17]

15 Karskens, 'Floods and floodmindedness', 325.
16 Van den Honert and McAneney, 'The 2011 Brisbane floods', 1151.
17 Middelmann et al., 'Riverine flood hazard', 102–06.

For all its size, Australia has a small population relative to comparable continents such as North America. At the point of British invasion and subsequent colonisation in 1788, the Indigenous population was an estimated 800,000, spread out in settings ranging from desert to sub-tropical coastal rainforest.[18] None of the Indigenous camps and settlements had large populations in particular localities, other than the much more closely settled Torres Strait Islands. However, all of the sites of the five settler cities were home to thriving Indigenous groups. These groups sustained themselves over millennia with the water available in their regions. They achieved this through detailed, intergenerational knowledge of Country and its patterns of water movement through and under the land, as well as respect for water as a living entity: water was and is held central to everyday material and spiritual life. The importance of water, and its power over life and death, is woven into legends that capture both the moral dimensions of water and its capacity to shape landscape over millennia. The changing seasonal abundance of water guided the location of campsites and seasonal patterns of movement between them: mobility was a key element of the groups' relationships with water. Indigenous water cultures vary, yet they share a common approach to water, based on understanding and respect. By contrast, settler approaches have usually been based on control, though recent consideration of water-sensitive urban design reflects a turn towards knowledge and adaptation.

European colonisers always sought out settlement prospects with ready fresh water, not just for drinking but also market gardening, given the necessity to be self-sufficient in fresh food as reliance on the irregular supply of dried foodstuffs from Britain was untenable. All early settlements had ocean or river access, with farming lands near to watercourses. Early industries included whaling and timber, with wool becoming the dominant export from the 1830s. Most long-distance movement of goods and people was by ship, whether around the continent itself or to and from Britain. Two of the five mainland capital cities, Sydney and Brisbane, were convict settlements; Perth received convicts from 1850 to 1868; Melbourne and Adelaide were free settler colonies. The prosperity of the wool industry, and the discovery of gold in Victoria and New South Wales in the 1850s, further integrated Melbourne and Sydney into a burgeoning world economy. By the end of the nineteenth century, both

[18] Hunter and Carmody, 'Estimating the Aboriginal population'.

TABLE 1.1. *Population of the five cities (thousands), 1851–2021*

Census date	Sydney	Brisbane	Perth	Melbourne	Adelaide
1851	54	3	8	29	18
1901	496	119	61	478	141
1947	1484	402	273	1226	382
1976	3144	1058	846	2764	940
2020	5367	2561	2125	5159	1377

Sources: McCarty, 'Australian capital cities in the nineteenth century', 119; Commonwealth of Australia, *Census.*

had close to half a million inhabitants, making them the 46th and 47th most populous cities in the world.[19] This extraordinary urban growth, in a country with a population at that time of a little less than 4 million, represents a pattern that continues to the present day (see Table 1.1).

The British crown claimed all of the Australian continent, including land and water, regarding the continent as '*terra nullius*' (nobody's land). This fiction was only overturned with the Mabo judgement in 1992, which acknowledged Native Title. British governors, as royal representatives, were installed in every colony (and continue to perform constitutional and ceremonial functions). In 1900, the British parliament gave permission for the Australian colonies to federate as the 'Commonwealth of Australia', with the colonies becoming states the following year. 'Crown land' still underpins land ownership in Australia, while the Crown owns the continent's subterranean resources, whether water or mineral. State governments oversee land sales, levying stamp duty on all land transfers. The primary landowners in every Australian state remain the state governments, which retain the right to appropriate private property for 'public goods', from freeways to dams, pipelines, and reservoirs.

Land grants and land sales were overseen by colonial governors and later Departments of Lands. Properties were subdivided for residential, commercial, and industrial purposes, while river and water reserves provided respite from the relentless march of privately owned land, enabling Aboriginal people to make semi-permanent homes. The most sought-after residential sites tended to look out over waterways, whether rivers, harbours, or later beaches; the wealthy preferred elevated sites, especially in flood-prone cities. Industries, especially those which needed plentiful water or ready waste disposal, also sought river sites; consequently,

[19] Chandler, *Four Thousand Years of Urban Growth*, 492.

tanneries, boat builders, brewers, flour mills, and paint manufactures were established along waterways. Later, the need for water supply prompted construction of coal-based gas plants and later electricity power stations near waterways.

Maps of the five cities in 2021, created by environmental designer Daniel Jan Martin and drawn to the same scale, show their oceanic settings, along with rivers, bays, and harbours. The current urban footprint is indicated in white shading, while protected areas are indicated by cross-hatched shading. These areas include national parks and nature reserves, preserved by state governments in perpetuity. Sydney is the only city of the five where protected areas serve as a key barrier to future urban growth. Melbourne has an almost unlimited hinterland for future urban development. Adelaide has partially regulated its urban sprawl, limiting development in the hill catchment areas to the east while allowing an ever-increasing north–south footprint. New suburban development in Perth and Brisbane heads to the north and south along the coast as well as moving inland along major transport corridors, following freeway systems.

Sydney (Map 1.1) developed on a semi-circular site, abutting the Pacific Ocean. It has substantial bays and river systems to its north (Broken Bay and the Hawkesbury River), west (the Nepean River), and south (Botany Bay and the Georges River). All three rivers flood after heavy rain, whereas the much deeper harbour, without riverine choke points, is not subject to flooding.

The initial British invasion force, predominantly composed of naval personnel and convicts, made their first settlement at Port Jackson (now Sydney Harbour), close to what is today Sydney's Central Business District, which offered a ready supply of fresh water. The Eora nation, traditional owners comprising a number of clans with their own territories, were decimated by introduced disease and frontier violence. The survivors' lifeways were devastated as successive governors, having claimed the continent for the Crown, handed out land grants to colonists for residential development and farms.

Much of the early settlement took place to the west and north-west of the town centre, with the valleys of the Parramatta, Nepean, and Hawkesbury Rivers all sought after for farming land. In the latter decades of the nineteenth century, suburban development followed train and tram routes. Reservoirs and water towers were built at high points around the growing metropolis. The water supply was augmented by five dams built in bushland to the south of the city between 1907 and 1941, with that

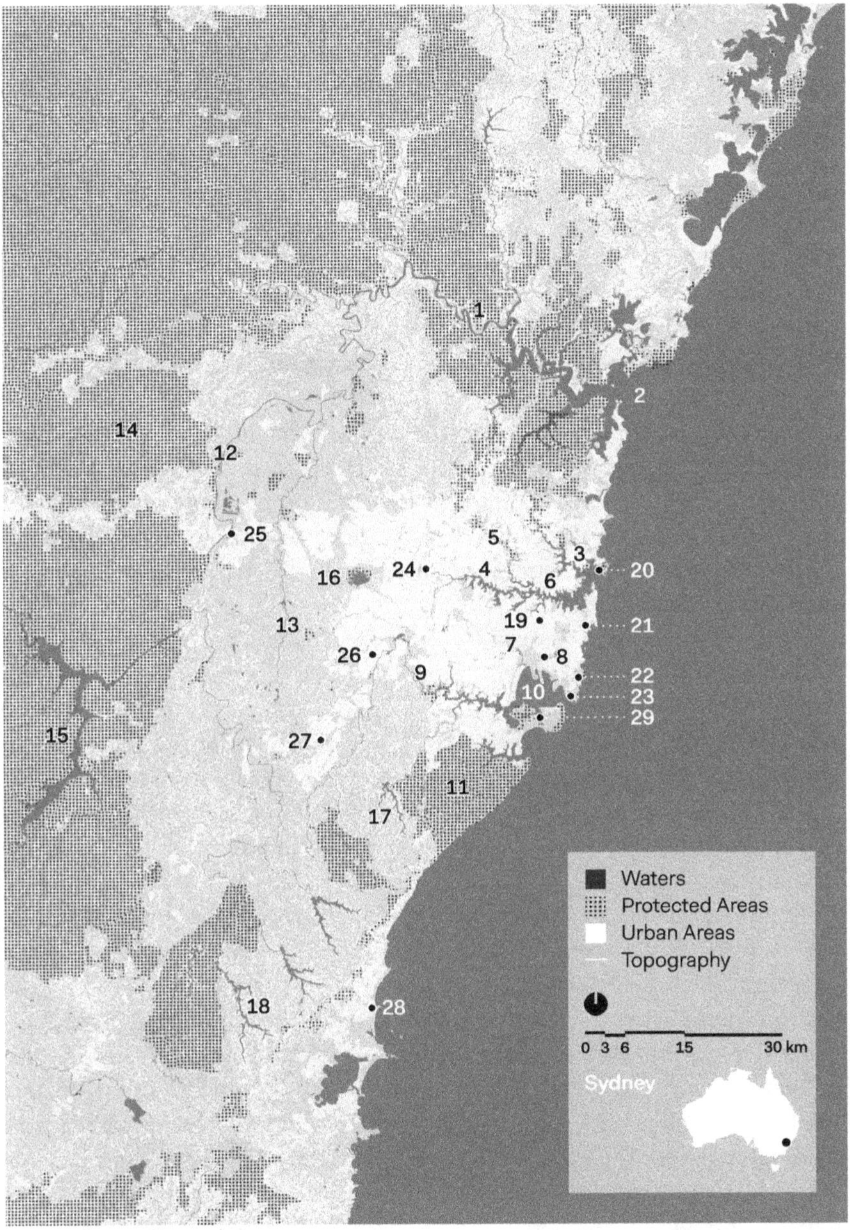

1 Hawkesbury River **2** Broken Bay **3** Middle Harbour **4** Parramatta River
5 Lane Cove River **6** Sydney Harbour **7** Cooks River **8** Botany Swamps
9 Georges River **10** Botany Bay **11** Royal National Park **12** Nepean River
13 Badgerys Creek **14** Blue Mountains National Park **15** Warragamba Dam
16 Prospect Reservoir **17** Lake Woronora **18** Lake Avon **19** City of Sydney
20 Manly **21** Bondi **22** Malabar **23** La Perouse **24** Parramatta **25** Penrith
26 Liverpool **27** Campbelltown **28** Wollongong **29** Kurnell Desalination Plant

MAP 1.1. Sydney (created by Daniel Jan Martin).

supply almost quadrupled by the Warragamba Dam, which opened in 1960. The creation of large national parks, Ku-ring-gai Chase to the north, National Park to the south, and the Blue Mountains to the west, curtailed suburban development, allowing Sydney more guaranteed green space and better-quality water catchment areas than any other Australian capital city. Sydney is often said to be blessed by its spectacular harbour and riverine topography, which made it feasible to build so many dams, also allowing cheap and easy solutions to the disposal of stormwater, sewage, and industrial waste, which created major pollution problems in the latter decades of the twentieth century. Public outcry eventually pressured the extension of the city's ocean outfall sewers further into the sea.

Between 1901 and 2021, Sydney's population grew tenfold, now standing at 5 million. Mass car ownership from the late 1950s saw suburban developments take over most remaining farming land. Since the 1970s, about one-third of all new housing has been provided by town houses and apartment blocks, at much higher densities than the traditional house and garden.

Brisbane developed around the Brisbane River, approximately 25 km upstream from its mouth in Moreton Bay (Map 1.2). The Brisbane River is 344 km long and within its 13,500 square kilometre catchment lies the sub-catchments of three rivers – Stanley, Brisbane, and Bremer Rivers – and twenty-two creeks, the largest of which is Lockyer Creek. The river is characterised by its meander, a propensity to flood, and its variable flow rate, which reflects the region's sub-tropical climate and high average rainfall.[20]

Meanjin, the lands of the Turrbal and Jagera peoples were occupied by the British in 1823 and named Brisbane. The British established the Moreton Bay penal colony in 1824 on the banks of Maiwar (the Turrbal name for the Brisbane River) and the current site of Brisbane's Central Business District, where the river offered a water supply and effluent dump. The penal colony closed in 1839 and the town opened for free settlement in 1842 as a part of the colony of New South Wales, gaining status as the colony of Queensland in 1859. Providing potable water was an early concern, solved by the construction of water tanks and the Enoggera Dam in 1866. Ongoing tension between managing water supply and flood mitigation culminated in the completion of two

[20] Only in southern Africa is the annual variability greater. Finlayson and McMahon, 'Australia V the World', 27.

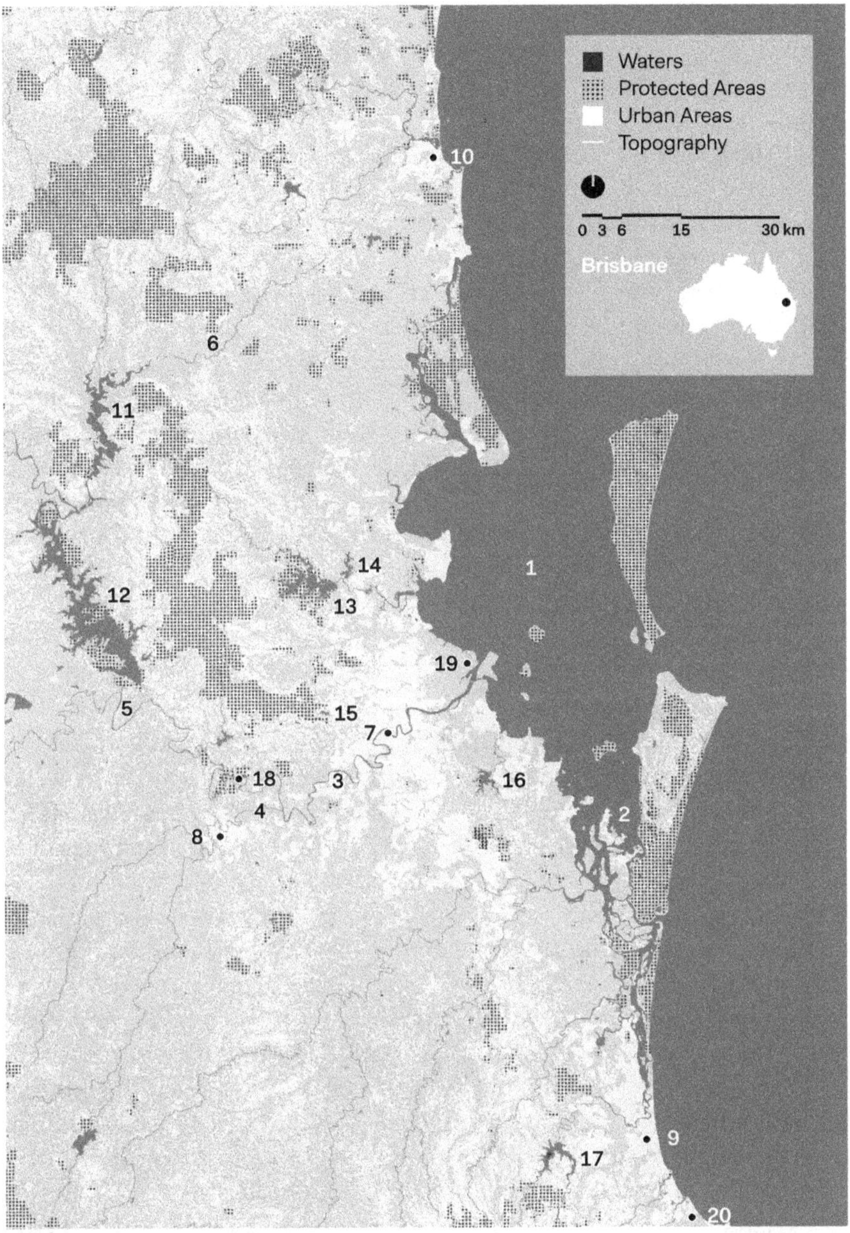

1 Moreton Bay 2 Stradbroke Islands 3 Brisbane River 4 Bremer River
5 Lockyer Creek 6 Stanley River 7 City of Brisbane 8 Ipswich
9 Gold Coast 10 Sunshine Coast 11 Somerset Dam 12 Wivenhoe Dam
13 North Pine Dam 14 Sideling Creek Dam 15 Enoggera Reservoir
16 Tingalpa Reservoir 17 Hinze Dam 18 Mt Crosby Wastewater Treatment Plant
19 Luggage Point Wastewater Treatment Plant 20 Tugun Desalination Plant

MAP 1.2. Brisbane (created by Daniel Jan Martin).

dual-purpose dams, Somerset Dam in 1959 and Wivenhoe Dam in 1984. Wivenhoe Dam provides most of Brisbane's water supply and, reflecting the region's variable climate, holds 2000 times the city's daily water supply when at full supply level.

In 1901, Brisbane's metropolitan area had a population of 119,000; in 2021, the number is fast approaching 2.5 million, making Brisbane Australia's third most populous city. The single house and garden on a 400 square metre block remain the dominant housing stock. The only capital city to adopt a metropolitan-wide council (in 1925), Brisbane spilled over into neighbouring local government areas in the 1970s. Since the 1990s, the delineation of the separate settlements on the Sunshine Coast, Gold Coast, Ipswich, and Brisbane has blurred, effectively creating a 200 km-long city.[21] Accordingly, water supply is managed as part of a complex grid that links dams, recycled water supply, and the Gold Coast desalination plant. With south-east Queensland's population growing by 80,000 people per year since 1981, the region's current population of 3.5 million is expected to reach 4.4 million by 2031 and 5.5 million by 2051.

The Perth metropolitan area is situated largely on a coastal plain that varies from around 23 km to 34 km in width, hemmed in by the hills of the Darling Scarp to the east and the Indian Ocean to the west (Map 1.3). Perth developed from Indigenous settlements established on the banks of the Derbarl Yerrigan, or what white settlers called the Swan River. While there was conflict with the Whadjuk Noongar almost from the beginning of colonisation, Aboriginal people have remained in and around Perth to the present day.

The colony's natural capacity to support European-style agriculture did not live up to the glowing reports of Captain James Stirling and botanist Charles Frazer, who reconnoitred the area in 1827. Much of the Swan Coastal Plain is characterised by sands with low levels of the nutrients required by European crops and livestock. The winters were cool and wet but the summers hot and dry, often with little significant rainfall falling over the summer months. With insufficient labour and no experience of farming in such an environment, the colony struggled until the introduction of British convicts in 1850. Modest growth followed between the port settlement at Fremantle, administrative centre in Perth, and the agricultural service centre of Guildford at the confluence of the

[21] Spearritt, 'The 200 km city'.

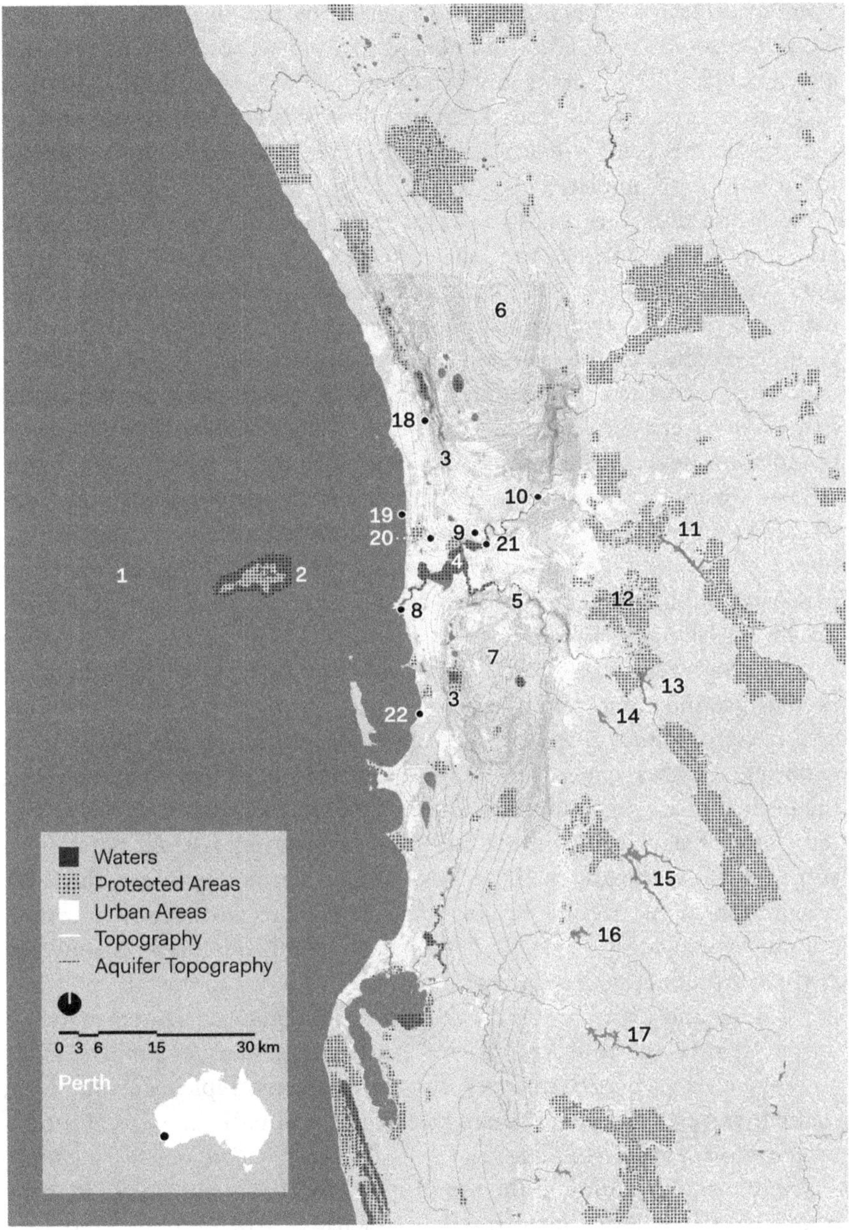

1 Indian Ocean **2** Wadjemup **3** Chains of Wetland Lakes **4** Swan River
5 Canning River **6** Gnangara Aquifer Mound **7** Jandakot Aquifer Mound
8 Fremantle **9** City of Perth **10** Guildford **11** Mundaring Weir
12 Victoria Reservoir **13** Canning Dam **14** Wungong Reservoir **15** Serpentine Dam
16 North Dandalup Dam **17** South Dandalup Dam **18** Groundwater Replenishment
19 City Beach Sewerage Outfall **20** Shenton Park Treatment Works
21 Old Burswood Sewerage Filter Beds **22** Kwinana Desalination Plant

MAP I.3. Perth (created by Daniel Jan Martin).

Swan and Helena Rivers. At first connected by the Swan River in 1883, when the settler population numbered around 8500, these nodes were also linked by Perth's first suburban railway. The forested hills to the east of Perth seemed the obvious location for water catchment and storage that could be gravity-fed to the city. The first such dam, Victoria Reservoir, was completed in 1891.

With the discovery in 1892–93 of large gold deposits at Coolgardie and Kalgoorlie, around 500 km east of Perth, the settler population quadrupled in a decade, with low-density suburban development extending around the railway line connecting Fremantle, Perth, and Guildford. The wealth generated by gold supported increased investment in public infrastructure, including a mains sewerage scheme to which the first homes were connected in 1911. After a short-lived experiment with treatment of sewage and disposal into the Swan River, which added to the effluent from noxious industries, Perth's first ocean outfall was constructed in 1926. But the low-density, distributed nature of urban development in a place with complex topography made centralised sewerage expensive, and pan toilets and increasingly septic tanks remained important means of waste disposal.

The boom of the 1960s and 1970s also saw increasing exploitation of Perth's groundwater resources, first to supplement the city's drinking water supply and then by householders sinking bores for garden watering. Further dams were built in the Darling Scarp, but following the decline in average rainfall in south-western Australia from the 1970s, the share of dam water in the city's drinking water fell. As population surged again in the 2000s in the wake of another mining boom, the city's orientation along around 80 km of coastline provided ready access to abundant seawater. The first desalination plant was commissioned at Kwinana in 2006, with desalinated seawater pumped to the hill reservoirs for storage and distribution; by 2019–20, desalinated seawater made up 43 per cent of the water supply for Perth's 2 million people. The city's aquifers also provided capacity for groundwater replenishment using treated wastewater, which commenced on a trial basis in 2010 and on a more permanent basis in 2017.

Melbourne's location at the top of Port Phillip Bay positioned it at the centre of the colony (later state) of Victoria, with a hinterland that included the major goldfields of the 1850s, rich pastoral land to the west, and the grasslands beyond the Great Dividing Range (Map 1.4). The gold discoveries of the 1850s pushed Melbourne past Sydney to become Australia's largest city and its financial capital. Sydney overtook

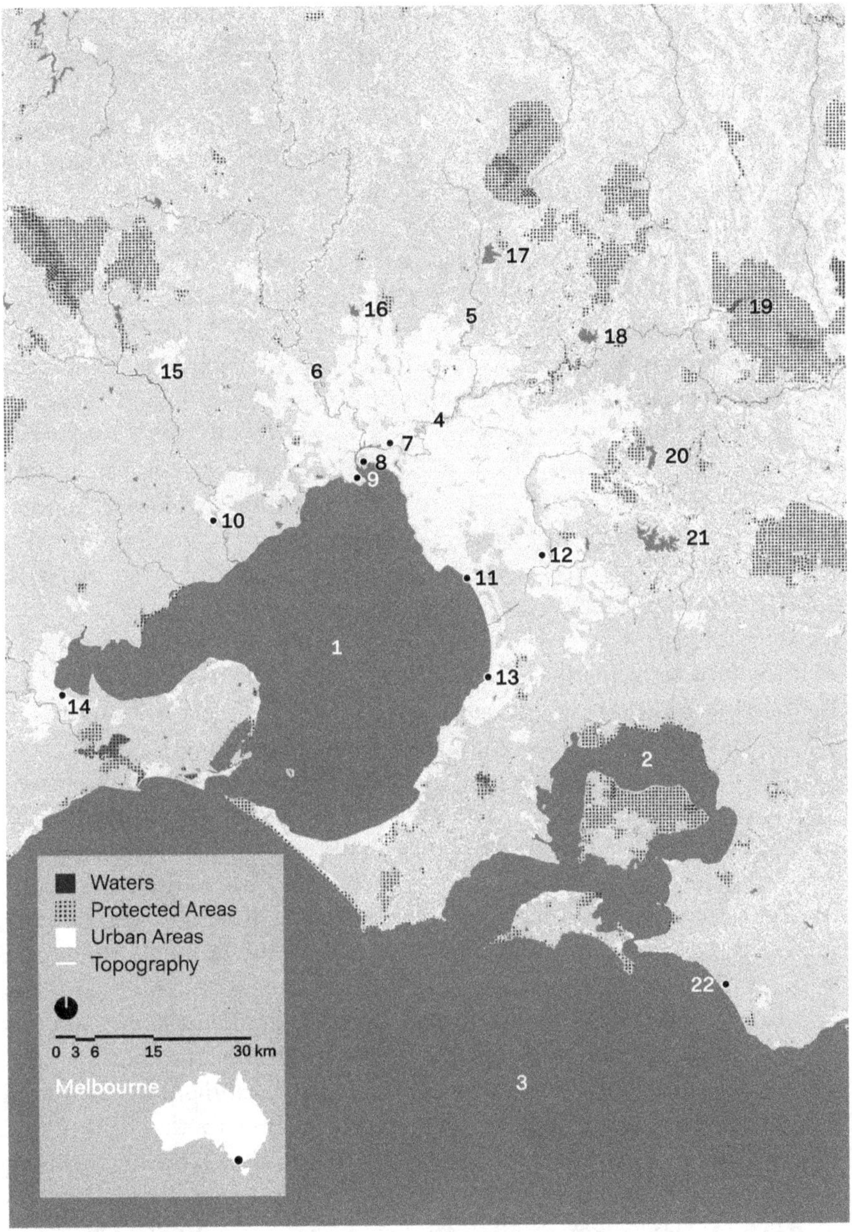

1 Port Phillip Bay 2 Western Port Bay 3 Bass Strait 4 Yarra River
5 Plenty River 6 Maribyrnong River 7 City of Melbourne 8 Port Melbourne
9 Williamstown 10 Werribee 11 Mordialloc 12 Dandenong
13 Frankston 14 Geelong 15 Melton Reservoir 16 Greenvale Reservoir
17 Yan Yean Reservoir 18 Sugarloaf Reservoir 19 Maroondah Reservoir
20 Silvan Reservoir 21 Cardinia Reservoir 22 Wonthaggi Desalination Plant

MAP I.4. Melbourne (created by Daniel Jan Martin).

Melbourne's population around the turn of the twentieth century, but had to wait another seven decades to become the nation's dominant financial centre.

The town began in 1835 as a settlement for squatters from Van Diemen's Land (Tasmania), who acted independently without government authority in acquiring land illegally from the Wurundjeri population. The site chosen was on elevated ground on the north bank of the Yarra River, where a natural rock ledge separated salt water from fresh, around 16 km from the river outlet at Hobson's Bay. Melburnians initially drew drinking water from the Yarra, but the river and its tributaries also served as sewers. The lack of strong tidal action in waterways and the Bay created difficulties in disposing of sewage. Along the 233 km course of the Yarra, floodplains and billabongs (oxbow lakes) hold water after rain and snow-melt; closer to Melbourne, the river snakes around natural obstacles before reaching an area of naturally swampy ground at the delta of the Yarra and Saltwater (Maribyrnong) rivers. Riverbanks were occupied by noxious trades that used waterways for waste disposal; poorly drained and flood-prone land nearby was occupied by manufacturing firms and worker housing. Further south, wetlands and lagoons were drained to form lakes, parkland, and building land.

Melbourne's sprawling, low-density built environment is a product of a preference for English-style suburban living, high average incomes, and government underwriting of suburban development through the provision of public transport. The separation of port towns (Port Melbourne and Williamstown) from the city centre made early railway connections a necessity; the city grid offered space for convenient rail termini, and wide, regular streets were conducive to the operation of trams. Middle-class suburbs extended south along the Bay, and to the east and south-east through pleasant hilly country and creek valleys extending to the Dandenong Ranges. To the north and west, the land was flatter and less inviting for suburban development but offered sites for industry and worker housing. South of the Dandenong Ranges, another large, flat area known as the Carrum Swamp fed creeks leading to the Bay. When drained, the Swamp would be turned into building land for industry and suburban housing.

When Adelaide was established in 1836 by the South Australian Company, the main settlement was centred on a string of waterholes known as Karrawirra Parri by the local Kaurna people and renamed the Torrens River. Colonel William Light's plan divided the town in two parts on either side of the river, each ringed by parklands. South

Adelaide would become the business and administrative centre, while over the next century outlying villages were gradually subsumed to become the greater metropolitan area. A separate port town, Port Adelaide, was laid out.

To Adelaide's east, the relatively low but grandly named Mt Lofty Ranges receive slightly higher annual falls (Map 1.5). Although as many as five streams now run from the ranges across the coastal plain, before white settlement they did not reach the sea, either running out of water or ending in swampy marsh areas. There are few perennial streams.

It took almost twenty-five years for the first reservoir to be completed, about 15 km along the Torrens, upstream from the city, while a basic sewerage system was in place by 1885. Apart from a boom in the late 1870s, the city population grew relatively slowly. However, water shortages were a regular occurrence. Despite seven major dams being constructed in the nearby hills between 1896 and 1977, before 1938 none had a capacity of 20,000 cubic metres. Only two, Mt Bold (1938) and South Para (1958), have a storage capacity over 30,000 cubic metres. To enhance water security, a pipeline to the Murray River, constructed in the late 1950s, now supplies 42 per cent Adelaide's water, although this can be as high as 90 per cent in a drought. As further security, a desalination plant, completed in 2013, was designed to supply up to 50 per cent of Adelaide's needs, although to date it supplies less than 2 per cent of annual consumption.

By 2021, approximately 1.3 million people lived in the greater metropolitan area of Adelaide – an urban landscape 20 km wide from east to west, and 100 km from north to south. Three major sewage plants discharge treated waste into Gulf St Vincent. The highly dispersed housing of Adelaide and suburbs has proven a challenge for infrastructure provision, although there is now a trend towards higher density living close by the city centre. Capturing and recycling storm water for drinking and reusing wastewater for gardens – both sources predicted to be essential for long-term sustainability – still only meet around 20 per cent of the city's needs. Most stormwater is discharged to the sea, although efforts to capture and reuse this are developing apace.[22]

In each of these cities, high average incomes, political stability, productive re-investment of the proceeds of resource extraction, and a preference for living in a suburban setting drove intervention in water supply

[22] Flynn, 'Designing a sustainable water industry'.

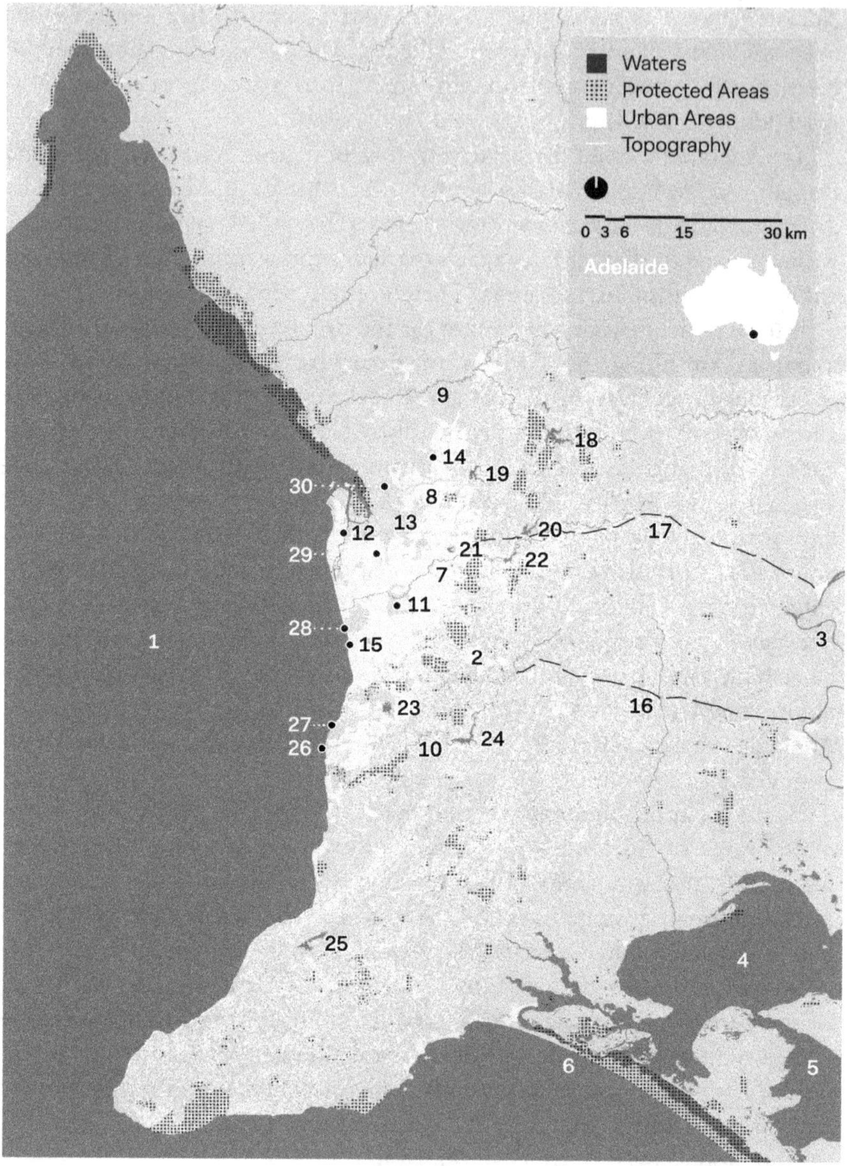

1 St Vincent Gulf **2** Mount Lofty Ranges **3** Murray River **4** Lake Alexandrina
5 Lower Lakes **6** Murray Mouth **7** Torrens River **8** Little Para River
9 Gawler River **10** Onkaparinga River **11** City of Adelaide **12** Port Adelaide
13 Salisbury **14** Elizabeth **15** Glenelg **16** Murray-Bridge – Onkaparinga Pipeline
17 Mannum – Adelaide Pipeline **18** South Para Reservoir **19** Little Para Reservoir
20 Millbrook Reservoir **21** Hope Valley Reservoir **22** Kangaroo Creek Reservoir
23 Happy Valley Reservoir **24** Mount Bold Reservoir **25** Myponga Reservoir
26 Adelaide Desalination Plant **27** Christies Beach Wastewater Treatment Plant
28 Glenelg Wastewater Treatment Plant **29** Islington Sewerage Works
30 Bolivar Wastewater Treatment Plant

MAP 1.5. Adelaide (created by Daniel Jan Martin).

infrastructure. In the twentieth century, heroic engineering solutions to water delivery and waste disposal were made possible by new construction methods and materials and government authorities with sufficient resources to pay for crucial capital works. State government bodies created the infrastructure, with high levels of private investment directed to detached dwellings in ever-growing suburbs. Housing standards for workers were comparable to those of most North American cities, but Australian workers could usually afford to rent or buy an entire house rather than share with other families. After 1890, when houses were more likely to have designated bathrooms, it became harder for public authorities to keep up with residential demand for water.

Aspirations for suburban living 'locked-in' a system of water movement over large distances, based on water availability, people, institutions, and legal instruments. Gravity and electric power kept dams, reservoirs, pipelines, sewerage mains, and wastewater treatment plants operating. Waterways were converted to impervious sewers, with natural creek lines drained and built over, limiting future options for storm water capture and storage. This contrasts with traditional Indigenous approaches to the availability of water sources, where in coastal areas with wetlands, rivers, and creeks small populations often relocated from one water source to another in alignment with patterns of seasonal change and abundance. These waterscapes were also valued for ceremony and sustenance, with Indigenous peoples intervening in water courses to channel eels and fish into traps and to water edible plants, as evidenced by the Gunditjmara system of *kooyang* (short-finned eel) aquaculture, established at least 6600 years ago in south-eastern Victoria and inscribed in 2019 on the UNESCO World Heritage List.[23]

At a global scale, the modernisation of urban infrastructure that provided greater control over water in the past two centuries is a component of what Deirdre McCloskey calls the 'Great Enrichment', an expansion in the range of goods and services to an average standard that was denied to the average person prior to 1800. In recent years, access to clean water (and electric washing machines) has reduced the burden of carrying water and handwashing clothes on poor women. The UN Millennium Development Goals (2000), which aimed to 'halve, by 2015, the proportion of people without sustainable access to safe drinking-water and basic sanitation', were achieved five years early.[24] In economist Robert

[23] Smith et al., 'Indigenous knowledge and resource management as World Heritage values'.
[24] McCloskey, *Bourgeois Equality*, 27.

Gordon's study of living standards in the United States after the Civil War, extension and expansion of infrastructure networks is central to improvements in individual well-being, through outcomes such as lower mortality rates, improved food quality and general health, improved housing standards, and increased investment in human capital. 'Networking inherently implies equality', Gordon concludes, as rich and poor eventually receive access to the same water, sewers, and drains, albeit after wealthy neighbourhoods enjoy an initial advantage.[25]

Detailed empirical studies of North American settler cities trace the development of what urban historian Martin V. Melosi calls the 'sanitary city' through the interaction of water supply and wastewater technology and the increasing sophistication of ongoing water management.[26] For cultural historical geographers Maria Kaika and Erik Swyngedouw, connecting people to network technology was a process 'infused by relations of power'.[27] In Western European, North American, and Australasian cities, nineteenth-century urban infrastructure such as water towers, reservoirs, and pumping stations were iconic local landmarks that 'provided the confirmation and lived experience that the road to a better society was under construction and paved with networks', but access to these benefits was distributed unequally, and the process was marked by exclusion and segregation.[28] In Australian cities, the wealthy generally gained access earlier than the poor. As the twentieth century progressed, government subsidisation of low-income and war service loans improved access to housing, and an internal bathroom became standard in most detached suburban houses built in the interwar years. But with the exception of the wealthy, lavatories located outside (colloquially known in Australia as 'dunnies') or tacked on to the rear of the house prevailed, particularly in older houses.

Australia's colonial and state governments provided the resources to meet the water needs of rapidly growing cities, yet in certain places and times infrastructure lagged or was provided in ways that imposed disproportionate costs on particular locations and their inhabitants. As Sydney drained untreated sewage into its harbour (with submarine outfalls later extended into the Pacific Ocean), low-lying areas of the city near the

[25] Gordon, *The Rise and Fall of American Growth*, 95.

[26] Melosi, *The Sanitary City*. See also Melosi, *Precious Commodity*; Tarr, *The Search for the Ultimate Sink*; Gandy, *Concrete and Clay*; Rawson, *Eden on the Charles*; Jacobson, *Ties That Bind*; Soll, *Empire of Water*.

[27] Kaika and Swyngedouw, 'The urbanization of nature', 98.

[28] Kaika and Swyngedouw, 'Fetishizing the modern city', 129.

outlets were subject to pollution. In Melbourne and Adelaide, citizens were willing to accept sewerage as a solution to public health crises, even though this entailed higher private costs than the continued use of cess-pits, pan toilets, and open drains. Australia serves as an example of the importance of political institutions, governance, and social and cultural determinants of water systems. Sustainable cities in the future are more likely to acknowledge the importance of social elements rather than disguising them within 'technological' responses.

In addressing urban water-related issues and crises, Australian cities responded in ways that were shaped by the current state of knowledge and perception of the nature of the problem. As cities grew, water supply and sanitation methods that were adequate in small settlements created environmental problems. In responding, colonial and local governments, private companies, and individual households pursued the easiest, lowest cost form of fresh water supply and wastewater disposal first, but the opportunity costs of such choices (the value of the next-best alternative use of resources) would rise over time. Like most aspects of the city-building process, the exploitation of water resources is a path-dependent process, highly sensitive to initial conditions that exert a lasting impact and shape and constrain subsequent land use.[29] Often the costs of decisions did not emerge for some decades. Differences in population and spatial size and structure meant that some cities could develop locally based solutions to crises, but by the twentieth century all five cities required solutions on a metropolitan scale. The Australian experience demonstrates that people were capable of changing how resources are used – at times rapidly. While change was not without cost and not without winners and losers, it was possible for large cities to be transformed into sites where human acquisition and transformation of water was reliable and abundant and met the expectations of residents and industries.

As these water systems developed, each city moved from simply tapping freshwater sources in largely ad hoc fashion to the systematic domesticating of water – shaping, regulating, and integrating water into networks of storages, pipes, and waste disposal facilities. Local government authorities proliferated in the latter half of the nineteenth century and early decades of the twentieth century. The larger cities followed British precedent and created statutory authorities charged with

[29] See David, 'Path dependence'.

responsibility for water supply, usually run by engineers, who were well informed about engineering advances in other parts of the world. These statutory authorities, or in the case of Brisbane its metropolitan-wide council formed in 1925, took over private and municipal water installations, including water towers and reservoirs. As they were able to levy water rates, these statutory authorities not only had their own relatively reliable income, they could also borrow money to build larger reservoirs and dams.

By 1901, Sydney and Melbourne were already so large that they could not continue to prosper unless their capacity to capture and store water was augmented. In all five cities, a strategy of establishing centralised management structures to build dams was pursued, despite variations in climate and local needs and despite water resources crossing State borders. This locked cities into a reliance on large, weather-dependent water storages that would inhibit change. Engineering solutions were responses to population growth that addressed supply-side issues rather than the demand for water, which was a product of cultural factors and habitually determined.[30] The adoption of Chadwickian water-flushed networks of sewage disposal created new demand for water, which led in turn to changes in the design of houses.

As the suburban frontier extended, water consumption, rather than conservation, became embedded in Australian culture, a process that interdisciplinary cultural researcher Zoë Sofoulis attributes to the rise of 'Big Water' – large-scaled centralised systems of single-use potable water provision.[31] Sofoulis coined the term during the Millennium Drought, as she critiqued the cultural and technical systems of Australian urban water management. Historically, settler water infrastructure for Australian cities has involved centralisation of water management, allowing for the concentration of political power and technocratic influence over environments, resources, expertise, and citizens that has persisted even with the corporatisation of the nation's water utilities since the 1980s. Such concentration of power can have undemocratic consequences and raises questions of transparency and accountability. When Sofoulis was writing, the cultural and technical systems of Big Water were also under challenge, albeit modest, as suburban households turned to (or revived) more independent, small-scale approaches to water supply and disposal, which

30 Leroux, Martin, and Zheng, 'Addressing water shortages by force of habit'.

31 Sofoulis, 'Big water, everyday water'.

indicated the potential for more decentralised and flexible forms of urban water management.

From the turn of the twentieth century, city dwellers came to expect a reliable water supply, but in times of drought only severe water restrictions could ensure that the cities would not run completely dry. These interruptions pierced the complacency afforded by largely invisible networks of water provision and disposal on which urban residents relied to develop and maintain their cultures of cleanliness, convenience, and civility both inside and outside their homes.[32] Australians who moved from the country to the city – as hundreds of thousands did in both the late nineteenth and twentieth centuries – often came from locations where tank water was the only form of supply, to cities where water was supplied, either into or to the rear of homes, at the turn of a tap.

City dwellers usually paid for their water either by a fixed charge per property or, in the latter decades of the twentieth century, based on the amount they used. In a supply-focused delivery system, water meters measured use in households, industries, and businesses. But water leaving the premises – for garden use, draining from kitchen, bathroom, laundry, and toilet – has never been consistently measured. Today most households receive a quarterly water and sewerage bill, based on the assumption that the water they buy to bring into the home is the same amount that goes back into the system as sewerage.

As curator and historian Jay Arthur observes, 'The colonist created [their] new country haunted by the image of the Default country, which was narrow, green, hilly and wet – which meant that *Australia* was understood as *vast, brown, flat* and *dry*'.[33] Settler Australia adopted a British framework for water management based on supply-side engineering solutions to moving water; by the late twentieth century, this approach was being challenged in response to drought, with an emphasis on 'demand management' solutions as well as localised rainwater capture, desalination, and the prospect of recycled water. Future climate pressures are likely to be much more demanding, and the legacy costs of past sprawl and current social and political settings (such as using markets to regulate demand and public uncertainty about scientific assurances that recycled water is safe) may make it more difficult for Australian cities to adopt more sustainable water systems at the required pace.

32 Head, 'Nature, networks and desire'. See also Kaika, 'Constructing scarcity'; Shove, *Comfort*; Kaika, *City of Flows*.

33 Arthur, 'The eighth day of creation', 73.

While Aboriginal people have not always enjoyed equal access to water in cities, Indigenous interests and rights in water were belatedly recognised in Australian national water policy in 2004; as cultural geographer Sue Jackson and Kamilaroi scientist Bradley Moggridge observe, this recognition 'provided an impetus for change in how we manage water, and the ways in which we seek knowledge of water, as well as appreciate its critical importance to all groups in Australian society'.[34] Although Indigenous efforts to secure the protection of urban wetland sites have sometimes been controversial, urban water utilities are increasingly seeking to acknowledge and engage with relevant Elders, and local Aboriginal water stories and cultures are becoming widely featured in public art and recognised in dual naming of rivers and other water sites. Settler Australia is belatedly realising that not only does justice demand recognition of Indigenous water rights, but also that there is much to learn from Indigenous ways of relating to water, developed over generations of living with Country.

By 2021, all of Australia's largest cities face problems that would be familiar to the residents of a century ago, albeit on a larger scale. While expecting reliable and safe water, their prosperity is threatened by the limitations of nature and the costs of maintaining, expanding, and improving existing and new networks. While new technologies, especially around water recycling, may enable a more sustainable water supply, such a dramatic change will be costly and can only happen if major political parties and voters support it, at a time when electorates are heterogeneous and public trust in governments is waning. The uneven impact of globalisation and technological change has made voters in middle and outer suburban, regional, and rural Australia more sensitive to job insecurity and cost of living pressures than those in well-off central and inner cities. Exemplifying the workings of the so-called hydro-illogical cycle, Australians are largely complacent about their real water usage and sustainability in this dry and drying continent with rapid urban development. This book aims to unsettle this uncritical approach and challenge these cultural assumptions.

[34] Jackson and Moggridge, 'Indigenous water management', 193.

2

Living Water

Water is life. All living organisms need water to survive. Water cycles through and connects the community of life on earth. This truth was, and is, deeply understood by Aboriginal people across Australia.[1] From the wet forests of southern Australia to the dry interior, and perhaps most of all in the resource-rich places that became Australia's capital cities, water occupied a central place in the physical and spiritual landscapes that Aboriginal people refer to as 'Country'. The diverse water systems and ecologies of these places sustained Aboriginal peoples for thousands of years because of two key factors: Aboriginal knowledge of water and associated wetland and riparian ecologies, and respect for life-sustaining water as a central tenet of Aboriginal cultures. As Whadjuk Noongar scholar Noel Nannup puts it:

> Water is connected to things you can't see – our spirituality, identity and survival. That's why water is sacred to us. It's alive, it gives us life. We consider water to be just as important as humans. It's what connects all living things.[2]

Australia's First Nations peoples are highly diverse. More than 250 distinct language groups with distinct histories, rituals, traditions, and economic strategies were thriving across the continent in 1788. While custodial responsibility for water, and some form of spiritual and cultural

[1] This book is principally a history, so in this chapter we often use past tense to refer to Aboriginal relations with water. This is not intended to imply that in the present such relations no longer exist or have fundamentally changed; in spite of the violent rupture of colonisation, Aboriginal ways of being and knowing have survived – and been revived – in many contexts across the continent.

[2] Nannup, 'Walk with the Waugal'.

connection to it, is common to all of these groups, their fine-grained water-related traditions and practices vary.[3] The ways in which the peoples of what are now the five cities used and understood water were developed and adapted over millennia of landscape change – including the often-violent and destructive changes wrought by the onset of colonisation. In a time of environmental crisis and ongoing ecological collapse, there is much for settler people to learn from the water practices of peoples who successfully sustained themselves with these lands for over 60,000 years.

SUSTAINING PHYSICAL AND SPIRITUAL LIFE

Water is a substance that flows through time: the water circulating in the present is the same water that has circulated since early in the history of the earth. The same water has passed through innumerable cycles of evaporation, rainfall, surface and sub-surface flows in and through bodies – human and non-human – and landscapes before evaporating and beginning the cycle again. One exception is water confined deep underground in aquifers like the Yarragadee, over which the Perth metropolitan area sits. Yarragadee water is mostly held in a layer of sand and porous rock more than 2 km thick; the youngest water in the aquifer is around 600 years old, the oldest more than 35,000. The Whadjuk Noongar people believe that this, the deepest of three aquifers beneath Perth, is home to the oldest *Waugal*, or rainbow serpent. In the words of Noel Nannup, the Yarragadee 'is the deep sleeping place of the *Waugal's wirn* [spirit], so you can understand why groundwater is incredibly important and sacred to us'.[4]

The chain of wetlands that run north–south along the length of the Swan Coastal Plain was made by the *Waugal*; they are also where its spirit went 'back in' and often still dwells today, in deep water and underground.[5] While the surface waters are most immediately connected with the unconfined superficial aquifer, the three aquifers – the superficial; the intermediate and partly confined Leederville aquifer; the deepest and mostly confined Yarragadee aquifer – all form an interconnected system. The Whadjuk Noongar have long understood that the surface waters and groundwater of the Swan Coastal Plain are connected: the chains of

[3] Moggridge and Thompson, 'Cultural value of water'.
[4] Nannup, 'Walk with the Waugal'.
[5] Goode and Associates, 'South west regional', 20; Wooltorton et al., 'Living water'.

wetlands that run between old sandy swales are where the groundwater reaches the surface. As Noongar elder Cedric Jacobs puts it: 'The water on the surface is really the marks where the *Waakal* [*Waugal*] either wound his way through or came up after making subterranean streams and waterways.'[6] The *Waugal* story preserves and explains this deep understanding, acquired through long and careful observation, of how water moves above, through, and under the land.[7] The *Waugal* has been described as 'a personification, or perhaps more correctly animalisation, of the vital force of running water', and therefore the question of whether any particular water source has (or belongs to, or is associated with a *Waugal*) 'becomes, from an Aboriginal viewpoint, meaningless and condescending. The presence of living water bespeaks Waugal immanence'.[8]

The enduring presence of these *Waugal* connects the past and present and has always demanded care and respect from the Noongar custodians, as water only continues to flow in the *Waugal*'s presence. As Whadjuk Noongar/Balardong oral historian Dorothy Winmar says of Noongar people:

> They believe in the *Waakal* [*Waugal*] very dearly. They reckon without the Waakal around they would have no water. They would not let the kids go and torment the *Waakal*. They (Noongar) would drive them away. There is a *Waakal* in the Swan River and he very rarely shows himself. If the water was muddy, the old grannies used to say don't swim in there, because he is having a feed. Don't swim (*warra wirrin*, or bad spirit); wait until the water is clear then you can go and jump in (*quop wirrin*, or good spirit). He was very important to their lives, because they believed in having fresh water. They wanted the water, so they wanted the snake to stay alive.[9]

While wetlands were/are commonly inhabited by *Waugal*, not all *Waugal* are the same. A dangerous *Waugal* inhabited Gnangara lake, so Noongar did not reside there.[10]

Rainbow serpents are associated with water across Aboriginal Australia. The present site of Brisbane, for example, is Turrbal and Jagera country.[11] There, one story of the creation of *Maiwar*/the Brisbane River tells that *Moodagurra*, the rainbow serpent, created the

[6] Jacobs, 'General interview', interviewed by Maxine Laurie, Town of Vincent Local History Centre, 8 January, 2008, in Bertram and Murphy, *In Time*, 161.
[7] Macintyre et al., *Report on an Ethnographic*, 23.
[8] O'Connor et al., *Report on an Investigation*, 47.
[9] Dorothy Winmar, interviewed by Len Collard, 2002, in Collard, 'Cosmology', 125.
[10] O'Connor et al., 'Report', 28.
[11] Steele, *Aboriginal Pathways*, 121–22. Turrbal is also spelt as Turbal and Turrubul.

land. One year, as she made her way up a dry creek, her large body became stuck. *Moodagurra* called on *Yara* (the rain) and *Ngalan* (the cloud) to help. As the storm thundered and rain flowed into the creek, the water seeped under the rainbow serpent's belly, allowing her to wriggle from side to side. As *Moodagurra* journeyed up the creek, her movement formed the sinuous curves of *Maiwar*.[12] Now *Moodagurra* brings the floods and decides when the big rains come. In a similar version, the river is said to be created by a fresh water eel, *Duelgum*, a totem of the Turrbal people.[13]

The significance of water to Aboriginal peoples is reflected in not only cosmology but also language. A settler account of Noongar words used in the early nineteenth century, George Fletcher Moore's *A Descriptive Vocabulary of the Language in Common Use Amongst the Aborigines of Western Australia*, contains no fewer than thirty-five Noongar words for types of water, in addition to several words for waterfowl, fish, shellfish, and aquatic plants. These 'water words' include an adjective, *kallăng*, for 'warm', which was applied only to water. This might be used to describe 'water standing in the whole of a rock, and therefore warm at any season under an Australian sun; water at the edges of lakes in the summer season'.[14] *Budulu* referred to both 'calm weather favourable for fishing' and also 'a space of smooth, glassy water', while '*gabbi yurro*' described 'the discoloured stream of fresh water, which descends after rain from the uplands mingling with the salt water in the estuaries'.[15] There were also numerous words for water in pools, lakes, streams, springs, wells, and oceans. Place names commonly contained references to water. For example, Beeliar, a chain of wetlands and now the name of the Perth suburb in which the remnant wetlands are located, means 'the run of water on a westerly direction by or through that place'.[16]

12 Ruska, 'Moodagurra'. Shannon Ruska is a storyteller and a Nunukul Yuggera man. This account was provided by the Brisbane Museum and the Dandiiri Library and Community Centre.

13 Turrbal and Wakka Wakka elder Joe Kirk described a Jagera variation serpent story, in which the Brisbane River is created by a fresh water eel, or Duegum, the spiritual totem of Jagera people. Uncle Joe Kirk, interviewed by Margaret Cook, 22 May 2017.

14 Moore, *Descriptive Vocabulary*, 53. Note that Moore used a phonetic transcription that does not always correspond to contemporary Noongar words.

15 Moore, *Descriptive Vocabulary*, 20, 116.

16 Collard, Leonard, et al., The Nyungar Boodjera Wangkiny = Nyungar land is speaking: Place nomenclature of the southwest of Western Australia project. www.boodjar.sis.uwa.edu.au/index.htm, accessed 25 March 2020.

The vocabulary of Kaurna words compiled by C.G. Teichelmann and C.W. Schürmann in early Adelaide similarly yields some specialised water words, such as *kauwemela*, for drizzling rain or other small drops of water, such as the whale blows out of its nostrils, and *yertalla*, 'water running by the side of a river'. *Kudnawodli* was recorded as referring to 'water closet, privy', suggesting that there were distinct places set aside for this purpose. Excrement, *kudna*, was seen as a powerful and dangerous substance, and its ability to contaminate water is reflected in Kaurna belief that river water may be 'enchanted' with *kudnamurro*, making the water dangerous to drink.[17]

In a place of long, hot summers, the Kaurna valued rivers highly. The *Karrawirra Parri*/Torrens River was a source of not only drinking water but also food and recreation. Early settler accounts describe how the part of the river that ran through what is now the parklands of central Adelaide included a chain of large waterholes, each 25–90 m apart, so deep that in some parts divers could not reach the bottom. These were surrounded by overhanging eucalypts and connected by small streams running over pebbles and through reeds and rushes that kept the water crystal clear for the interdependent life around the river. Freshwater crayfish and mussels were abundant in the larger deep holes, and Kaurna women were 'adept at diving to get them, staying under water a considerable time and coming to the surface with perhaps five or six in their hands or in any way they could conveniently carry them'.[18]

In Whadjuk Noongar country, the chains of wetlands running along the coastal plain were crucial for subsistence, providing a source of not only fresh water but also fibre and food: bulrush, frogs, turtles, marron, waterbirds, and eggs.[19] The wetlands also attracted game: kangaroos, emus, and bandicoots. They were therefore popular camp sites, particularly during the summer months. In winter, when parts of the coastal plain would have become inundated and difficult to traverse, families moved eastward along watercourses and up into the Darling Range, where they hunted game and harvested yams. In spring, the smaller groups would come together into larger groups and camp by the flourishing freshwater places of the Swan Coastal Plain.[20] Changes in surface

17 Teichelmann and Schürmann, 'Outlines'.

18 J. Chittleborough, 'Primitive Adelaide: Recollections and impressions', *Register* (Adelaide), 27 December 1906.

19 Meagher, 'The food resources', 14–65.

20 O'Connor et al., *Report on an Investigation*, 15.

waters were an integral part of the seasonal rhythms of Whadjuk Noongar life.

Like the Whadjuk Noongar, the Turrbal and Jagera of what is now Brisbane moved with the seasons, finding abundance and avoiding depletion of any particular resource. As Jagera man Neville Bonner explained in 1995, 'we rotated around allowing nature to provide and crops to rejuvenate.'[21] Inland people gathered in large camps near lakes and rivers during the drier, cooler season, living off the abundant resources in the lowland open forests and wetlands. They dispersed in the warmer wet season, when water was more widely available across the landscape, to forage and hunt in smaller family groups.[22] *Maiwar*, an integral part of the spiritual landscape, also provided a bountiful source of seasonal food including mullet, flounder, crabs, shellfish, turtles, eels, and waterfowl.

Campsites in Turrbal and Jagera country were functionally permanent, if not occupied at all times, and carefully designed. There were different areas for young men, young women, families, menstruating women, and pregnant women, as well as areas for dancing and meeting. There were separate toilet areas, and for campsites located on a river, drinking areas upstream of washing areas. Heavy items such as bark sheets and grinding stones were stored on departure for use on return. Campsites could accommodate anywhere from 20 to 200 people – or more for major feasts and corroborees.[23] They were located in places with good access to a range of amenities: water for drinking, food, and recreation were principal considerations. One camp site was in *Kurilpa* or place of the water rats, a triangle of land on *Maiwar* (now the site of the suburbs of West End, South Brisbane, and Highgate Hill). Its waterholes, creeks, and wetlands offered both a rich array of flora and fauna and a shallow river crossing at low tide.[24] Nearby Woolloongabba was *Woolloon-camemm* (whirling water), a creek with a series of waterholes that in rain would run from one to another, swirling around one hole before rushing to the

[21] Neville Bonner, interviewed by Robyn Buchanan, 1995, https://pictureipswich.com.au/nodes/view/2164#idx5339, accessed 18 April 2021.

[22] SKM & Aurecon, 'Cross River rail report: Indigenous cultural heritage', July 2011, http://eisdocs.dsdip.qld.gov.au/Completed%20Projects/Cross%20River%20Rail/EIS/EIS%2030%20Aug%202011/01%20Volume%201/18%20Indigenous%20Cultural%20Heritage.pdf, accessed 12 April 2021.

[23] Kerkhove, *Aboriginal Campsites*, 1–5.

[24] Kerkhove, *Aboriginal Campsites*, 81; Joe Kirk discussion; Clarke, 'A Jubilee retrospect', *Queenslander*, 7 August 1909; Petrie, *Tom Petrie's*, 89, 148.

next.[25] Settler William Clark wrote in 1849 that One-Mile Swamp in Woolloongabba was a camping ground, and a large waterhole (where the Brisbane Cricket Ground – the 'Gabba' – now stands) was a favourite swimming hole for both Aboriginal and settler people prior to the 1840s, when it was largely filled in.

Brisbane's extensive creek system also offered many camp sites. Settler accounts of the city describe Kedron Brook as teeming with fish (mullet, eel, and catfish), and the nearby flats at Kalinga were a large corroboree site. Smaller campsites included *Nundah* (from Nanda chain of water-holes) and nearby *Nudgee* (place of ducks), and the Boondall wetlands provided water and food supplies.[26]

Maiwar and its tributaries were also used as for transportation. In deeper areas the Turrbal and Jagera used rafts and canoes or swam, holding logs for flotation when fatigued. Reminiscing in 1909 about South Brisbane in the 1840s, settler William Clark recalled Turrbal people crossing the river on rafts or up to sixty people swimming across at a time, holding spears above their heads.[27]

In the region that is now Sydney, the often-rugged topography meant that rivers formed crucial transport and communication lines. Country, people, and languages were familiar and connected along the length of the river systems, but not beyond. By 1788, the Gweagal, Bediagal, and Cobragal people's societies were 'canoe cultures'. All of their members were adept at constructing canoes, though they might be used differently according to gender. Women commonly fished with hooks from canoes, men with spears from rocks.[28] These practices and technologies were not, however, evidence of a timeless culture; the *burra* (shell fish hooks) were a major innovation that emerged between 500 and 1200 years before the present.[29]

Noongar people traversed their territory on well-defined and established routes along and between fresh water sites. Map 2.1, created by Daniel Martin, shows the *Derbarl Yerrigan*/Swan River and some of the wetlands remaining to the present day, such as *Galup*/Lake Monger and *Boodjamooling*/Hyde Park, as well as wetlands (dark grey) and the extent of natural floods and flows (lighter grey) now eradicated by the suburbs of

25 Steele, *Aboriginal Pathways*, 133. 26 Kerkhove, *Aboriginal Campsites*, 55.
27 Petrie, *Tom*, 89–91; Clarke, 'A Jubilee retrospect', 21.
28 Karskens, *The Colony*, 386–447; Goodall and Cadzow, *Rivers and Resilience*, 33.
29 Attenbrow, *Sydney's Aboriginal Past*, 18–21. On coastal Aboriginal people's fishing culture and its endurance beyond 1788, see Irish, *Hidden in Plain View*.

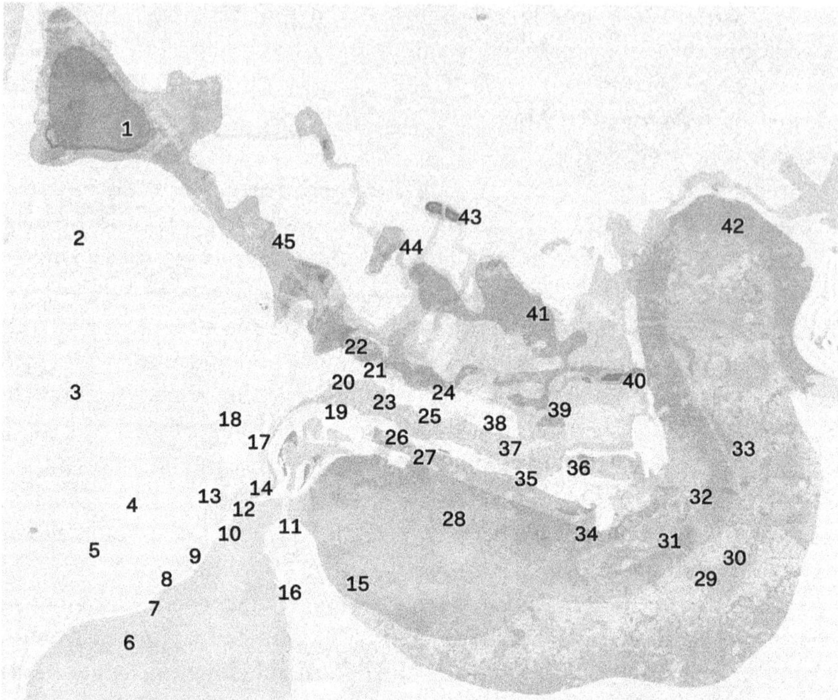

1 Galup
2 Goobabilup
3 Wandaraguttagurrup
4 Gargatup
5 Gabbee
6 Goodroo
7 Goomap
8 Goonialup
9 Kalgardaup
10 Kooyamulyup
11 Kareenup
12 Goonininup
13 Karrakatta
14 Kata Moor
15 Gaboordjoolup
16 Booryulup
17 Gabbee Kalga
18 Byerbup
19 Goodinup
20 Byererup
21 Gooloogoolup
22 Biabulup
23 Karboordup
24 Yandilup
25 Kuraree
26 Gumap
27 Gooyagarup
28 Buneenboro
29 Mindeera
30 Joorolup
31 Goonagar
32 Kakaroomup
33 Yoondoorup
34 Matta Gerup
35 Dyeedyallalup
36 Mardalup
37 Kureejup
38 Widodootogup
39 Weitch-Rutta
40 Goongoongup
41 Yoordgoorading
42 Warndulier
43 Boodjamooling
44 Boojoormelup
45 Goologoolup

MAP 2.1. Derbarl Yerrigan/Swan River and some of the wetlands remaining to the present day. The wetlands are indicated in dark grey and extent of natural floods and flows in light grey. Many of the wetlands have now been eradicated by the suburbs of Perth (shown in overlay) (created by Daniel Jan Martin).

Perth (shown in overlay). Based on research by Whadjuk Noongar scholar Len Collard, among others, the map highlights the historical extent of wetland around what is now the Central Business District and its significance to Noongar people, reflected in the density of naming.[30] Whadjuk Noongar places all existed in a web of temporal as well as spatial relation to each other, and this might be reflected in their names. The name of Lake Goollelal (*gool – lel – al*) means 'to travel on or by or hereabouts', reflecting its status as the first major camping ground on the route between *Galup*/Lake Monger ('place of home or the hearth') and the important ritual and corroboree area at Yanchep.[31] Even after colonisation, as the settlers drained and filled the wetlands around the centre of what is now Perth, Whadjuk Noongar woman Fanny Balbuk famously continued to walk her *bidi* (track connecting water places) between *Matta Gerup* ('shin-deep' – a crossing point on the *Derbarl Yerrigan*) and *Gooloogoolup*/Lake Kingsford (destroyed to build Perth Railway Station). Journalist Daisy Bates described how Balbuk would climb or break through fences erected across the *bidi*, even forcing her way through a house that was built over it.[32]

While enjoying its benefits and accepting the responsibilities of stewardship, Aboriginal people were well aware of the risks associated with living water. On Noongar country, fear of the *Waugal* kept children away from deep water in which they might drown. Across Australia, floods were appreciated as an essential part of the river's life-cycle that shaped country, created floodplains, and replenished life – both human and non-human – in a river's catchment.[33] However, floods were also understood as a risk. In the greater Brisbane area, Turrbal and Jagera camps were built under trees to provide shade and near water, but on sloping ground at least 14 m above water courses.[34] In 1890, the Cooyar people told journalist Archibald Meston about a large flood on the Upper Brisbane River, which they called *Magenjie*, Big Flowing Water or Big River.[35] In 1842, Aboriginal people warned the McConnel family, new settlers in the

[30] Collard and Palmer, Noonookurt Nyininy; Department of Local Government, Sport and Cultural Industries, Gnarla Boodja Mili Mili. www.dlgsc.wa.gov.au/department/publications/publication/gnarla-boodja-mili-mili-(our-country-on-paper), accessed 12 April 2021.

[31] Wooltorton et al., 'Living water', 13.

[32] Bates, *The Passing*, 70.

[33] Goodall, 'The river runs', 41.

[34] Kerkhove, *Aboriginal Campsites*, 1.

[35] Meston, 'Floods and droughts', *Queenslander*, 29 March 1890. The meaning 'big flowing water' for Magenjie is given by C. Moynihan, 'To the editor', *Queenslander*, 17 August 1901.

Brisbane River Valley, that an inundation had occurred the previous year.[36] Meteorologist Inigo Jones wrote in 1929 that Aboriginal tradition said Brisbane floods could originate in the Stanley River.[37] Settlers chose to build on the *Maiwar* flood plain, putting themselves in harm's way.

The recollections of Thomas Petrie, who from 1837 grew up among the Turrbal and learned their language and customs, provide a settler perspective on how mid-nineteenth-century Turrbal people responded to scarcity of water. Like flood, drought also had its advantages. When river water levels were low, eels could be caught by 'men muddying the water, and feeling for them with their feet'.[38] When water was scarce, the Turrbal would dig small wells in swamps; they then filled the wells with fern leaves to clarify the water, making the sediment sink. Water might be carried from place to place in a container called a *pikki*, made from bark or the sheath of the bangalow palm flower.[39] A *Turrwan* or 'great man' was believed to have power to bring or stop rain. For example, a leader of the North Pine people called Dalaipi was regarded as the rainmaker for their Country.[40] In Petrie's account, rainmaking was said to be linked to crystal stones or *kundri*, found where the spirit of the rainbow touched water.[41] As Marcia Langton, herself a descendant of the Iman people of Queensland, puts it, for Aboriginal people 'water and water places are constructed socially through the experiences of the knowledgeable men and women who live in these landscapes as tenure systems and spiritscapes'.[42]

WATER AND TERRITORY

> Noongars are pattern-thinkers and cycle-watchers. We remember the last Ice Age, still tell stories of the Cold Times. Deep Time is a stone dropped in a still pond and Noongars read the ripples. This Country remembers what it was. It remembers everything that has ever been. Settler Time overwhelmed us, but Deep Time endures.[43]

For thousands of years, water moved across and beneath the land, and people moved between water places in daily and seasonal rhythms. But

[36] Elizaberth Coxen, 'Notes on floods in the Brisbane River', Read before the Royal Society of Queensland, 21 August 1893. BCA 1577.

[37] 'Flood prevention', *Queensland Times* (Ipswich), 2 November 1929.

[38] Petrie, *Reminiscences*, 74. [39] Petrie, *Reminiscences*, 94, 105.

[40] Petrie, *Reminiscences*, 29, 185–86. [41] Petrie, *Reminiscences*, 29.

[42] Langton, 'Earth', 290. [43] Lynch, 'Split', 8.

water also shaped mobility over much longer time-scales. Aboriginal people and their ancestors have occupied the sites of Australia's metropolitan areas since at least 40,000 years ago and have thus lived through dramatic landscape changes. At that time, the sea level was around 90 m lower than at present. It retreated further at the Last Glacial Maximum around 20,000 years ago, when sea levels were 120 m below present. From there, as the polar ice receded, sea levels rose at a rate of around 1 m per century, with two periods of rapid rise, at 15,000–13,000 years ago and 11,000–9000 years ago, when sea levels rose up to 2.5 m per century before stabilising around 6000 years ago.[44] Depending on the topography, inundation could occur very quickly: on gently sloping country in northern Australia, more than 1.5 km of land could be lost every year during the period of most rapid rise.[45] Within just 12,000 years, the landmass of greater Australia shrunk by around 23 per cent; around 85 per cent of that land was occupied immediately prior to inundation.[46]

This experience of coastal inundation and remaking of homelands is captured in Aboriginal tradition from many areas around Australia. Noongar tradition, as told to colonist George Fletcher Moore, is that Rottnest, Carnac, and Garden Islands, now 5–18 km offshore from metropolitan Perth, were once part of the mainland. One day an intense fire burned across the landscape and the ground split; the sea then rushed in between the islands and the mainland.[47] This tradition dates from at least 8900–7500 years ago. Similarly, the Boon Wurrung and Woiworrung people told early colonists at Hobson's Bay (within Port Phillip Bay, now embraced by metropolitan Melbourne) that the Yarra River 'once went out at the Heads, but that the sea broke in and that Hobson's Bay, which was once a hunting ground, became what it is'.[48] Another tradition about the creation of the Yarra River describes how the water of the Yarra used to be locked in the mountains, but powerful ancestors, Bar-wool and Yan-yan, cut channels through the hills, creating the Plenty and Yarra rivers and releasing the water onto the plain now called Port Phillip Bay.[49] Traditions from the area also recount hunting on the grounds across what is now the Bay and the rushing in of water

44 Murray-Wallace, 'Pleistocene', 717; Stanford et al., 'Sea-level', 202.
45 Hanebuth et al., 'Rapid flooding', 1033; see also Flood, *The Original.*
46 Gautney and Holliday, 'New estimations', 103–12.
47 Nunn and Reid, 'Aboriginal memories', 30.
48 Hull, 'Report', quoted in Nunn and Reid, 'Aboriginal memories', 18.
49 Massola, *Bunjil's Cave*, 58.

during a catastrophic event with strong winds and subsidence. It is unclear whether these stories refer to the initial inundation of Port Phillip Bay around 9000 years ago or a more recent event around 2800–1000 years ago when the entrance to the bay was blocked with sediment and the plain subsequently re-inundated.[50]

When confronted with sea-level rise, the ancestral Noongar, Boon Wurrung, Woiworrung, and other Australian Aboriginal peoples retreated to higher ground. This process must have involved protracted and perhaps difficult negotiations and adjustments for cultures in which, as Patrick Nunn and Nick Reid put it, people are

> consubstantial with species and places; where people are known to country and where country's health is dependent on people's actions; where people's lives are formed and their well-being sustained through knowing country; and where belonging arises through knowing to the extent that people feel like strangers off-country.[51]

These attributes also, however, ensured that knowledge of inundation and retreat was embedded within cultural tradition in a way that saw it preserved for thousands of years. From studying ancient responses to sea-level rise, including episodes from north-western Europe in the more recent past, Nunn and Reed conclude that retreat was a more successful strategy than the construction of defences against sea-level rise. These old stories, which have endured across many generations, suggest that contemporary urban dwellers are better off working with Country, with long time-scales in mind. This would entail developing robust long-term plans to live with the rising sea levels resulting from climate breakdown, rather than attempting to preserve current coastal investments through short-term, piecemeal adaptations based on wishful thinking and a reluctance to change. This includes, of course, planning for resilient fresh water supplies; at present, much of Australia's desalination and wastewater treatment infrastructure lies on vulnerable coastlines.

URBAN FRONTIERS

When the colonisers arrived, they often looked, consciously or otherwise, to Aboriginal settlements as the most strategic, resource-rich, and pleasant areas to appropriate for their own settlements. While some of the

[50] Nunn and Reid, 'Aboriginal memories', 18.
[51] Nunn and Reid, 'Aboriginal memories', 41.

advantages of the places in which Aboriginal people routinely camped were natural, others were a result of careful land management by Aboriginal custodians over generations. As historian Ray Kerkhove points out, John Oxley and Alan Cunningham favoured Breakfast Creek as the site for a Moreton Bay colony because it was 'numerously inhabited' by Aboriginal people, and the explorers understood that Aboriginal camps indicated rich land and plentiful water.[52] Wynnum, today an eastern suburb of Brisbane, was an Aboriginal camp from which hunters launched expeditions to neighbouring islands in search of dugong, turtle, and flying fox. Settlers followed suit, and as their town grew it engulfed the camp sites, which were then turned into public parks. Redcliffe, now a north-eastern suburb of Brisbane, was the site of Queensland's first European settlement. Matthew Flinders recorded an Aboriginal camp there in 1799, though it was not in use at the time. This was a place for fishing mullet and mud crab, collecting oysters and hunting turtle and dugong, but it also had abundant and reliable water in the form of lagoons, swamps, and wells. The colonists also focused on fish, oysters, and crabs, while using the Aboriginal wells. The old well sites became water reserves, persisting to the present as public open space. Kerkhove identifies several other instances, including at Nambour, Maroochydore, and Ipswich, where significant Aboriginal residential areas pre-dated the European ones, and settlers began exploiting the same resources that had drawn Aboriginal people to the area.

As in the American West, most settlers saw towns as agents of civilisation, with urban progress representing a moral triumph over the 'barbarism' and 'savagery' of 'wilderness' and its inhabitants.[53] In competing to attract migrants and investment, settler city-builders celebrated progress in buying and selling town lots, often while holding a negative – or at least instrumental – view of nature. 'This district', wrote the editor of the *Moreton Bay Courier* in 1846, 'which so very recently echoed only to the foot-fall of the wandering savage, now exhibits all the signs of civilisation'.[54] In the early days of colonisation, local Indigenous people rarely sought to settle in town, preferring to live on and with the land as they always had, in communal ways. As white and Indigenous lives intersected, the newcomers' understandings of the purpose of land led to conflict with the existing landholders.

52 Kerkhove, 'Aboriginal camps'.
53 Hamer, *New Towns in the New World*, 206–19.
54 'Labour', *Moreton Bay Courier* (Brisbane), 4 July 1846.

Australian colonial frontiers were therefore not only pastoral: as historian Penelope Edmonds puts it, 'early towns and cities were developing urban frontiers, equally charged and often-violent contact zones of racialised spatial contestations'.[55] Edmonds puts forward the town camps as important '"other spaces" within the nineteenth-century array of new sites created by the processes of settler colonialism'.[56] Where town officials promoted settler colonial cities as 'spaces of civility, progress, control and racial purity', the reality was more complex: Aboriginal people ventured forth from the camps into the towns, and whites came to the camps seeking entertainment, trade, or interracial sex. These contact zones were inevitably located on wetlands or river banks that had not been enclosed as private property.

The Colonial Office in London was concerned about the issue of Aboriginal welfare and protection and recommended that South Australia's founding Act be amended to include a recognition of prior Aboriginal occupancy. This amendment was not made prior to settlement but was included in the Letters Patent and Orders-in-Council, the legal instruments authorising the colony's establishment. A Protector of Aborigines was appointed, but did not arrive in the colony until 1839.[57] By that time, the Kaurna had been moved to a nearby 'Native Location' between Adelaide and Thebarton, settlers having reserved their original camp site on the *Karrawirra Parri*/Torrens River as the access point for water carters and the location for a slaughterhouse. Within ten years of settlement, all Kaurna land and water resources on the Adelaide Plains were expropriated by Europeans.[58] Kaurna remained close to the river, camping in the city's parklands, where their presence attracted complaints from settlers about noise, immodesty, and even their continued use of the river for bathing and recreation. By 1840, the Editors of the *Register* opined: 'Something ... ought certainly to be done to keep the Natives out of the river. From the Botanic Garden, downwards, it is full of them; not much, we should think, to the improvement of the water, which all the inhabitants who have not wells are obliged to use'.[59] Before long, settlers turned to violence in order to prevent Kaurna use of the river. In 1843, Adelaide resident W. A. Crowthorne noted in his diary: 'a policeman came with orders to burn all their wurlies or huts which was done ... the

[55] Edmonds, 'The intimate', 131. [56] Edmonds, 'The intimate', 143.
[57] Foster and Nettlebeck, 'From Protectorate', 27–28.
[58] Lockwood, 'Early encounters', 66–68. [59] *Adelaide Register*, 15 February 1840.

reason for them burning them out was because they swam and made the water dirty above the [water] hole'.[60]

Perhaps more critical to Kaurna dispossession was the undermining of their traditional economy by colonisers who understood little of how to manage the river and cared less. The depth of the waterholes on the *Karrawirra Parri* had meant that water was available all summer, even when the river was not running. But the colonisers cut the trees and tea-tree for firewood and building materials and took the pebbles and gravel over which the streams between pools had run for building and road-making. An early settler recalled in 1906 the changes he had witnessed:

> when the natural surface of the stream was destroyed the banks between the large pools were gradually washed away and deposited in the bottom of the deep holes. The process continued until the river assumed its present desolate appearance where the water is not held back by artificial means. How different is the Torrens now from what it was in those early days![61]

In Brisbane, in the years immediately after colonisation, Aboriginal camps were located near water crossings, including Enoggera and Breakfast Creek.[62] From the 1840s, Enoggera Creek and Enoggera Crossing were used as a border: Aboriginal people were not permitted to cross into the main colonial settlement and were punished by police if they did so.[63] For seventy years, Aboriginal people and settlers, including Chinese market gardeners, co-existed in this region, although there are recorded incidents of violence between 1847 and 1857. Construction of Enoggera Dam in 1866 significantly reduced the water and game available to Aboriginal people living in the area. Corroborees were still conducted at Enoggera Camp up to the 1890s, although with declining numbers. The size of the camp site was reduced by encroaching suburban land development after 1901.[64]

Lagoons and creeklets in *Barrambin* (York's Hollow) in central Brisbane drained into Breakfast Creek, just downstream from Enoggera Creek. Breakfast Creek was one of the largest and most important Aboriginal camping areas in the region, with several sites. Sand islands made for convenient river crossings. The river had deep lagoons, and the Turrbal people built fish traps to catch the plentiful mullet, whiting, bream, jewfish, and eels. After colonisation, until around the 1860s,

[60] Cited by Jones, 'Kaurna cultural heritage', 140.
[61] Chittleborough, 'Primitive Adelaide'.
[62] Kerkhove, *Aboriginal Campsites*; Joe Kirk, discussion.
[63] Fisher, 'From depredation', 43. [64] Kerkhove, *Aboriginal Campsites*, 71–74.

Aboriginal people supplied much of the Brisbane market for fish from this traditional fishery. Like Enoggera Creek, the place also marked a boundary between settler and Aboriginal space, with Aboriginal people being driven out of town across the creek every evening. Aboriginal people were hostile to any whites entering their side for over thirty years; the Indigenous community to this day regards it as a battle line.[65] The camps remained in place until the 1880s, even though they were burned to the ground by police in 1852, 1860, 1861, 1865, and 1867. Aboriginal people returned as the camps' locations on the periphery of the main settlements offered not only diminishing riparian resources and memories of happier times but also opportunities for employment, trade, spiritual sustenance, and continuation of customary duties.[66]

Rivers also remained crucial places for Aboriginal people in Sydney. Shortly before Governor King's departure from New South Wales in 1806, a deputation of Darug representatives told him:

> they did not like being driven from the few places left on the banks of the river where alone they could procure food ... if they went across white man's grounds the settlers fired upon them ... that if they could retain some places on the lower parts of the river they would be satisfied. ... Their request appears to be so just and equitable I assured them that no more settlements would made lower down the river.

As Langton dryly notes, subsequent governors soon forgot these reassurances.[67] For other Aboriginal people in the region, rivers provided rapid mobility as well as the rugged gully terrain essential to the guerrilla war they were waging in the form of attacks on settler crops.[68] As colonial occupation spread in the growing town and beyond, the increasingly urbanised rivers remained home to local people from Gadigal, Darug, Dharawal, and other groups as well as people displaced from country further afield. Aboriginal cultural processes of place-making in one locality could be applied to another: such processes could 'fulfil those central principles of Aboriginal society that are enabled by relationships to country: that is, to take on a mature role in the social network by fulfilling responsibilities to look after land'.[69]

In Melbourne, where fenced properties covered hunting lands and fishing waters became fouled by drains carrying storm water, sewage, and industrial waste, the Aboriginal Protectorate system aimed to remove

[65] Kerkhove, *Aboriginal Campsites*, 82–83.
[66] Kerkhove, *Aboriginal Campsites*, 85.
[67] Langton, 'They made a solitude', 37.
[68] Goodall and Cadzow, *Rivers*, 44–47.
[69] Goodall and Cadzow, *Rivers*, 17.

Woiworrung and other Kulin Nation people from the streets of Melbourne. A mission was established on the south bank of the Yarra, beside a 'lush swamp'.[70] The residents frequently moved between the mission and fledgling town, as well as their chosen campsites near the mission reserve land and in the north-east and north-west parts of town.[71] As more Kulin land was engulfed by the growing city, as in other cities, areas reserved for water – wetlands, river reserves, water reserves – escaped the march of private property and provided refuge for Aboriginal people. Spaces along Melbourne's smaller watercourses were 'transitional within the process of colonization; they were nervous spaces that were not yet property'; wetlands perhaps more robustly resisted conversion into commodified space, especially given that settlers commonly associated them with disease and death.[72] Although the settler city wished to render Aboriginal people invisible, it still relied on their labour into the 1840s for many basic services, including water carrying.[73]

Today, most of the Aboriginal heritage sites in the Perth metropolitan area are associated with rivers, lakes, and wetlands.[74] Aboriginal people continued to live in these sites beyond colonisation: *Galbamaanup*/Lake Claremont was a popular campsite for Noongar people until the 1950s, when the camp was demolished by the local council.[75] In the late 1980s, a Noongar man born in the 1910s recalled Lake Claremont as 'a paradise' due to its abundance of fresh water, possums, turtles, snakes, and waterfowl.[76] *Boodjamooling*, now highly manicured Hyde Park with two ornamental lakes at its heart, was another major campsite and meeting place for Noongars from the Swan Coastal Plain and beyond.

In the wake of colonisation, Aboriginal people did not disappear from the sites of the growing cities. Though many people were killed by disease or violence or forcibly relocated to reserves or missions, some continued to live in and around their traditional campsites – all of which were associated with water sources. Historian Paul Irish and Gadhungal researcher Michael Ingrey observe that Aboriginal camps in urbanising areas are widely regarded as undesirable sites not of Aboriginal people's choosing.[77] While this was sometimes the case, Aboriginal people also continued to live in and maintain the areas with good water and other

[70] Edmonds, 'The intimate', 87. [71] Edmonds, 'The intimate', 122.
[72] Edmonds, 'The intimate', 133, 143. [73] Edmonds, 'The intimate', 138.
[74] O'Connor et al., 'Report on an investigation', 12. [75] Cook, *That Was My Home*.
[76] O'Connor et al., 'Report on an investigation', 21.
[77] Irish and Ingrey, 'Marking their footsteps', 84.

resources that they had in some instances occupied for thousands of years. Rich riparian environments supported Aboriginal resilience before and through the trauma and deprivations of colonisation, providing 'an intensively harvested economic source, a rapid transport and communication system, and a source of cultural symbol and narrative'.[78] This pattern of informal Aboriginal occupation of watery places persisted well into the twentieth century, even as urban rivers and wetlands became degraded and polluted.

Wetlands and river sites remain a key focus of Aboriginal struggles for respect and cultural recognition and have featured extensively in Aboriginal resistance and revival movements in the various capital cities. By the 1930s, Salt Pan Creek, a tributary of Sydney's Georges River, was home to a long-standing camp that had nurtured Aboriginal activists from all over the state and beyond.[79] Musgrave Park, the 'unofficial political capital of Aboriginal Brisbane', has a long history as an Aboriginal residential area and water source.[80] In 1978, a group of Aboriginal men caught turtles in Perth's Hyde Park lakes and lit a fire on which to cook them; they were moved on by police for disorderly behaviour.[81]

A TIME FOR LEARNING

Aboriginal societies across Australia respected, understood, and valued water in the landscape; this relationship enriched, sustained, and often saved their lives. Over long periods of living with Country, Aboriginal groups witnessed – and adapted to – significant changes in climate, landforms, flora, and fauna. This meant that, as historians Heather Goodall and Alison Cadzow put it, 'Aboriginal traditional societies were the custodians of flexible, responsive cultures which allowed Aboriginal people both to keep the best and richest of their traditions and yet still record and cope with changes.'[82] These cultural principles and processes endured in spite of the shock and devastation of colonisation. The diverse Aboriginal peoples across Australia had developed water cultures appropriate for their particular part of the continent, yet some broad water-related principles appear common to all. One is knowledge: understanding the affordances and risks of different forms of water and preserving

78 Goodall and Cadzow, *Rivers*, 8.
79 Goodall and Cadzow, *Rivers*, 1–4.
80 Kerkhove, 'Aboriginal camps'.
81 O'Connor et al., 'Report on an investigation', 33.
82 Goodall and Cadzow, *Rivers*, 12.

these in robust oral traditions. Another related principle is mobility: following the water and its seasonal abundances and avoiding its seasonal hazards. Some cultures also adopted water as a fast and reliable means of transport that bound communities together across large tracts of Country.

For all Indigenous cultures, water is also a storied medium that connects the past and present in the 'long now': a living and lively substance that sustains the Country of Whadjuk Noongar, Turrbal, Kaurna, Boon Wurrung, Woiworrung, Gadigal, Darug, Dharawal, and other people. Indigenous approaches to water are no longer dominant in Australia, having been displaced by more control-oriented settler approaches. There is much work to be done in order to ensure that Aboriginal peoples' rights and interests in urban water are recognised, but there are some encouraging signs of settler attention to Aboriginal voices. Even within what are now Australia's most densely populated areas, Aboriginal ways of apprehending water sites 'as places in a cosmology can be used by all visitors and residents, to nurture a sense of respect, care and familiarity.'[83]

[83] Wooltorton et al., 'Living Water', 5.

3

Domesticating Water

'Water is procurable at a reasonable depth, and the natural slope of the ground provides for ample drainage', promised an 1896 poster advertising Glebe estate, overlooking Perth's Swan River at Freshwater Bay.[1] As the entry point and service centre for the Eastern Goldfields at Kalgoorlie, Perth was growing fast, and in response the Anglican Church subdivided land it owned to create Glebe estate, in quarter-acre lots with 66-foot frontages and 12-foot-wide lanes at the rear. The son of one of the original buyers remembered a made road extending only to the first block. 'The rest was just grey sand and bush; and I rather feel sorry sometimes for our parents when they had to lug us up through that grey sand. There was no water, light, power where we lived – nothing.'[2] Labourer George Caporn and his wife Eliza lived in a hessian shanty in their backyard near Glebe estate while building their own three-bedroom weatherboard cottage, with a lean-to kitchen and a single tap at the back verandah. 'We never had a bathroom', son Roy recalled. Eight children 'used to bath in a big tub; we used to take it into our room'. Eliza 'used to do all her washing under a bit of cover outside, used to boil the clothes in two kerosene tins over an open fire'.[3]

Founded by the British as the administrative centre of the Swan River Colony in 1829, Perth was the last of the five mainland Australian capitals to grow into a major city. After exploring the area in 1827, Captain James Stirling reported that the 'supply of fresh water from

[1] Cited by Bolton and Gregory, *Claremont*, 61.
[2] Cited by Bolton and Gregory, *Claremont*, 67.
[3] Cited by Bolton and Gregory, *Claremont*, 67.

springs and lagoons is abundant . . . on the whole it may be confidently assumed that water is plentiful all over this territory'.[4] As population grew, natural water sources were subject to competing demands. As late as 1881, Perth had fewer than ten thousand inhabitants; the city then boomed due to a series of gold discoveries, with the population increasing to 107,000 by 1911.[5] Prior to the *Heath Act* of 1898, an absence of regulation regarding the drawing of water also applied to its disposal, through the building of cesspits (most of them unbricked), the dumping or burying of night soil on public land, or the draining of wastes to the Swan River. The sandy soil of the coastal plain was both a source of groundwater and a receptacle for waste water. Much of the area to the east and north of the early settlement of Perth was originally wetland; after draining, the high water table presented problems of damp and the contamination of underground water. Household actions imposed reciprocal costs: households polluted their own water supply and that of their neighbours. In creating and then dealing with these challenges, Perth's experience in domesticating its water was similar to that faced in the other cities.

'Domesticated water' refers to the process whereby water sources, such as rivers and lakes, and natural features, including streams, low-lying land, and the sea, are gradually shaped, regulated, and integrated into a system of dams, piped networks, and waste disposal facilities. Technologies that allowed individuals to obtain water from wells and waterways and use private cesspits to hold wastes were effective in small, low-density settlements but ill-adapted to large cities.[6] A shift from decentralised, local solutions to centralised, capital-intensive infrastructure changed the urban environment but also removed the individual from direct responsibility for water management. Instead of gathering and transporting water individually, water was supplied into and the waste disposed of from the homes of users.

This chapter considers the domestication of water in the five cities, a transformation that took more than a century and was driven by fear of epidemics, changes to understanding of disease, and the sanitation and engineering requirements associated with networked, high-density populations. We begin with an account of the impact of white settlement and urban demand for water on the Indigenous population. This is followed by a discussion of the institutional framework and governance structures that allowed citizens in settler colonial societies to express preferences for

[4] Quoted by Hunt, *Water*, ix. [5] McCarty, 'Australian capital cities', 136.
[6] Melosi, *The Sanitary City*, 11–15.

improved and safer water infrastructure. The chapter then considers how investment in water supply and sewage disposal created lasting effects that continued to shape infrastructure and water use in the twentieth century.

Clean water is a common resource, as defined by economists, when there is no restriction on its use and each use of the resource makes less available to others. In an influential 1968 essay, ecologist Garrett Hardin used an allegory of freely accessible common land, with humans making independent decisions about its use to maximise immediate payoffs and to predict overharvesting and eventual exhaustion of resources.[7] Political economist Elinor Ostrom challenged Hardin's conceptualisation of the 'tragedy of the commons', not on the grounds that the allegory was inappropriate or that common resources could not be overused, but with regard to the conclusion that the tragedy was one that would trap users due to their inability to self-organise and develop institutions to take collective action to prevent the problem.[8] Hardin saw only two possible solutions, each involving top-down action from governments: direct management or regulation of resource use through command and control policies, or the establishment of private property rights to facilitate efficient market allocation of resource access and use. Drawing on economic experiments and field work, Ostrom demonstrated that users of common resources were capable of developing their own management rules and productive norms of behaviour, without the need for a central authority to impose sanctions or offer incentives. Ostrom's was a theory of polycentric governance, suggesting that citizens may be empowered to develop institutions that are conducive to socially optimal resource use by the settings and contexts in which dialogue and collective action take place. 'What the research on social dilemmas demonstrates is a world of possibility rather than of necessity', Ostrom concludes. 'We are neither trapped in inexorable tragedies nor free of moral responsibility for creating and

[7] Hardin, 'The tragedy of the commons'. Hardin used the tragedy of the commons allegory to advance a neo-Malthusian argument for coerced population control and recognition of limits to carrying capacity, which was echoed in population policies in several countries in the mid-twentieth century. Hardin's case failed to take into account the possibility of falling birth rates worldwide associated with the demographic transition and the substitutability of factors of production in response to resource depletion. See Ridley, *The Rational Optimist*, 202–07; Hulme, *Why We Disagree About Climate Change*, 258–63.

[8] Ostrom, 'A behavioral approach'; Ostrom, 'Collective action'.

sustaining incentives that facilitate our own achievement of mutually productive outcomes'.[9]

Where the kind of institutions Ostrom describes fail to develop, water use is marked by conflict rather than cooperation. In the early stages of the five cities' development, people used water in ways that exacerbated scarcity, which manifested in environmental degradation and disease – in Perth's case, diminishing water quality and consequent epidemics of typhoid in the 1890s. Transmitted through contaminated water or food and by flies, especially in hot weather, typhoid posed the most serious threat to adult health in the late nineteenth century. The disease affected all age groups, but the fatality rate for adolescents and young adults was very high. Those who contracted typhoid and survived suffered permanent damage and would remain vulnerable to cardiovascular and respiratory disease.[10] In an era when a typhoid death rate greater than 20 per hundred thousand inhabitants was regarded as a sign of polluted water supplies, the 47 largest US cities averaged 58.[11] Melbourne's typhoid death rate averaged 75 from 1871 to 1880 and remained high during the 1880s, peaking at 126 during an 1889 epidemic. In Sydney, the rate averaged 83 from 1882 to 1886, peaking at 102 in 1885. Adelaide's inner city typhoid rates peaked at 209 in 1866.[12] Perth's typhoid death statistics are incomplete but are of a different order of magnitude to its contemporaries: in 1897, the peak epidemic year, typhoid death rates for the City of Perth reached 414.[13] Rapid population growth and inadequate provision for waste disposal created a common resource problem that necessitated finding local solutions.

When sufficiently large, networked water infrastructure provides safe water and effective sanitation, creating a form of public good: freely available, without one person's use noticeably reducing the resources available to others. By the end of the nineteenth century, Perth's population had quadrupled during the gold rush, and it became apparent that the city had made little progress in meeting the challenge of ensuring

9 Ostrom, 'A behavioral approach', 16.

10 Ferrie and Troesken, 'Water and Chicago's mortality transition'.

11 Galishoff, 'Triumph and failure'; United States Bureau of the Census, *Mortality Statistics*, 24.

12 Frost, 'Water technology', 24–27; Coward, *Out of Sight*, 127; Woodruff, *Two Million South Australians*, 37–42.

13 Calculated from data in Whittington, *Gold and Typhoid*, 414–15; Fraser, *Western Australian Year-book for 1896–97*, 42. The figure is inflated by the presence of hospitals in the City of Perth, to which people would have travelled to receive treatment from outside the city; a more accurate figure would require analysis of death certificates that record the home address of the deceased.

reliable water supplies. Perth lagged behind other Australian cities that had matured earlier. In 1810, the only city of the five to exist was Sydney, still a penal colony, with a population of six thousand (Hobart was founded as a second penal colony in 1804). A century later, over 1.6 million people lived in the five cities, three-quarters of them in Sydney or Melbourne.[14] Sydney's closest natural water sources were soon exhausted, prompting short-term solutions that required modification and improvement; outlets for sewage and storm water were built to local waterways. Facing massive gold rush population pressures, and with unpredictable rainfall ranging from drought to violent thunderstorm, Melbourne built its water supply system expediently, overcoming disputes between local governments, but was late in starting work on a sewerage system. Adelaide experienced slower population growth but faced the driest environment, drawing reticulated water from a reservoir on the Torrens from 1862. After years of procrastination, the city committed itself to a high-quality wastewater system connected to a sewage farm. Brisbane's small population limited the resources available for public investment in water infrastructure, which resulted in the construction of a series of dams across local creeks. The process of network improvement had a long gestation period and was neither universal nor linear, with cities leading in some areas while lagging in others.

The transition from localised, labour-intensive water movement to city-wide, capital-intensive networks was more than a matter of legislation and government spending: it required changes to society's attitudes and expectations and the embedding of these in legal codes.[15] Everyday use of water technology forged political connections between private and public life, encouraging a new civic dynamic to emerge with a politics of entitlement and provision in times of scarcity, as historians Vanessa Taylor and Frank Trentmann argue for Victorian London and Sheffield.[16] In Europe and North America, the rise of germ theory from the late 1870s and its broad acceptance by the 1890s shaped popular understanding of what made people sick and what kept them healthy, prompting private households to modify and exchange information about domestic practices and in turn exert pressure on authorities to provide cheap and abundant water and effective waste disposal.[17] This was not a solely top-down process but rather one that reflected the ability of citizens

[14] McCarty, 'Australian capital cities', 129–33. [15] Greif and Mokyr, 'Cognitive rules'.
[16] Taylor and Trentmann, 'Liquid politics'.
[17] Mokyr and Stein, 'Science, health, and household technology', 183–84.

in liberal democracies to take collective action to assert their rights to higher levels of quality in water infrastructure service, along the lines that Ostrom suggests.

Institutions – humanly devised rules and constraints that shape human interaction – may produce adverse economic and social outcomes that persist due to path-dependent processes, as advantaged individuals and organisations with bargaining power have a stake in perpetuating the system.[18] Urban built environments are also subject to path-dependence, due to the lasting impact of initial conditions that shape and constrain subsequent land use and choices of infrastructure. Governments were slow to adapt the form of cities to changing economic conditions and environmental challenges, as settlement sites, configurations of streets and public spaces, patterns of land ownership, and the costs of removing and replacing buildings and pipes acted as frictions to the renewal of urban systems. In periods of rapid population growth, the shortcomings of existing systems are accentuated, while the claims of existing residents are frequently entrenched. Unless a city suffers a major shock that necessitates complete rebuilding, planners and builders never deal with a blank slate.[19] Initial patterns become 'locked in' because low-cost solutions to urban problems are subject to increasing opportunity costs as populations increase, technology changes, and new environmental problems emerge. This is not a deterministic process but rather one in which political, cultural, economic, and technological variables affecting the city-building process are mediated by the lasting effects of institutions, geography, town siting, planning, and the placement of streets, buildings, and pipes.

The foundation of Australia's capital cities had tragic consequences, as settlers brought disease and violence to the Indigenous population and transformed their traditional lands and waterways. The growth of towns set in motion forces that led to the marginalisation of local Indigenous people and the extension of the frontier along inland rivers and grasslands. For settler invaders, town prosperity provided motivation and resources to attempt to bring natural water resources under their control. The engineering technology adopted for water supply, drainage, sewerage, flood control, and land filling was already in use overseas, but public investment in infrastructure tended to be sparing and delayed until it was decided which level of government should be responsible. The provision

[18] North, *Transaction Costs, Institutions, and Economic Performance*. See also David, 'Path dependence'.
[19] Siolda, 'Razing San Francisco'.

of urban amenities was affected by weak planning powers and skills and difficulties in creating effective administrative controls and overcoming vested interests and tax resistance.[20] By the time investment was required, the cities had grown beyond the original corporation boundaries. Later residential segregation and fragmented local government reflected diverse environmental conditions across metropolitan areas. Dispersed urban areas raised the costs of infrastructure provision, and new suburbs were usually built without proper drainage and street paving. On flat land close to rivers, industry demanded water supplies for production and used waterways for waste disposal, conditions that workers were willing to support because of the link between local jobs and low taxes and production costs.[21] As the nineteenth century proceeded, more inclusive political institutions and aspirations for better living standards were conducive to civic engagement and citizen-based problem solving. New knowledge about causes of disease altered the willingness of households to trade off environmental amenity to reduce cost of living pressures, allowing incremental changes to take place.

WATER: LIFE, DISPLACEMENT, AND DEATH

In establishing colonies in Australia, Britain assumed the land was *terra nullius* and located towns on sites that Indigenous people were still using in large numbers. As historian Grace Karskens observes, 'the British were a maritime people – water was their familiar and preferred mode of travel, and rivers were seen as conduits into the unknown interior'.[22] They sought bases along coastlines where natural contours of the land created harbours, where waterways could be used for transport, water supply, and waste disposal, and where river valleys provided food, timber, and animal resources. These settlements would become gateways to prosperous hinterlands – hubs where produce was shifted between various forms of land transport and coastal or oceangoing ships, where commercial services were provided, farm inputs were produced, and rural output was processed. Like other European colonial powers, the British not only brought disease and guns but also changed the context in which the lives of Indigenous people unfolded through the control of water for city building. As shown in Chapter 2, the beginning of the dispossession of

[20] Ergas and Pincus, 'Infrastructure and colonial socialism', 230.
[21] See for example Lack, *A History of Footscray*.
[22] Karskens, 'Floods and flood-mindedness', 319.

local Indigenous people was marked by a reshaping of the Australian landscape to meet urban demand for water, fuel, and construction materials by clearing vegetation, altering watercourses, and draining and filling wetlands.

In each of the five cities, the siting of early townships failed to account for environmental hazards. Brisbane, Melbourne, and Perth were built on flood plains; Adelaide and Sydney had areas that were low-lying and poorly drained, with risks of localised flooding and pollution from drain outlets. Despite evidence that indicated a high risk of flooding, settlers in port cities and inland towns were willing to buy or occupy low-lying land and to build and live on it, as they did on the Hawkesbury-Nepean Rivers. As Karskens writes, for settlers,

> the floods were powerful and malevolent entities. They were understood in terms of wildness, they were wilful, unbridled, impetuous, they surged far beyond puny human ideas of order and control. The risks of death and ruination were integral to a kind of contract they had with the river.[23]

The choices people made to settle and build on flood plains, highly risky from a modern perspective, were comparable to those of contemporary migrants from British rural areas who headed for industrial cities, where the risk of death was higher. Some were lured by the prospect of higher wages, which compensated for local environmental hazards. Others saw the potential payoffs of being close to where regular work could be found. Others were forward-looking, seeing better long-run prospects than in rural areas.[24] In Melbourne and Manchester, Brisbane and Bradford, urban locations offered potential benefits that for many outweighed the associated risks. In settler cities, some rainfall prior to white settlement was collected by Aboriginal people in 'water trees' or *gnammas* (rock holes), but most was absorbed into subsoil, used by plants, evaporated, or drained into waterways. Urbanisation slowly altered this water cycle as settlers cleared vegetation and created impervious surfaces, such as house roofs and paved streets, that increased runoff and erosion. Clearing land and building close to waterways exacerbated flood damage, polluted water sources, and degraded the immediate surrounds. These interventions contributed to social issues, as classes became segregated by the distribution of environmental, social, and lifestyle assets and risks within

[23] Karskens, 'Floods and flood-mindedness', 337. See also Cook, *A River with a City Problem*, 7–16; O'Gorman, *Flood Country*, 15–59.

[24] Mokyr, *The Enlightened Economy*, 455–57. See also Williamson, *Coping with City Growth*.

particular urban contexts. Elevated sites were typically well served with infrastructure and occupied by elite housing, with cheaper flat land, often the earlier sites of Aboriginal camps and fishing places, used for industry and worker housing.

Proximity to fresh water was a key influence on the siting of the five cities. In every case except Perth, the first site chosen for settlement in the region, or the natural harbour site, was rejected for lack of water. Arriving in 1788 with a thousand or so convicts and soldiers, Governor Phillip found Kamay (Botany Bay) lacking in fresh water and moved his headquarters to Port Jackson, where the large, deep harbour offered better shelter for ships. A freshwater stream, originating in swampland west of present-day Hyde Park, determined the location of the settlement at Sydney Cove. In 1823, New South Wales Surveyor General John Oxley, tasked by Governor Brisbane with finding a site for a new penal colony, located a waterway at Moreton Bay that he named the Brisbane River.[25] The settlement at Redcliffe was abandoned after one year, due to poor water supplies and lack of a suitable port. The new site chosen for what became Brisbane was 27 km up the river, where there were freshwater creeks. Still navigable, the Brisbane River surrounded the town on three sides, making it an ideal location for a prison. The penal settlement was closed in 1839, and in preparation for the arrival of free settlers, surveyor Robert Dixon laid out a grid of 20 m-wide streets to control and administer land subdivision. To capitalise on the demand for land, the New South Wales surveyor general rescinded an earlier order that all land within 100 feet of the river's high-water mark be reserved for open space. Despite floods in 1841, 1843, and 1844, demand for riverside land was strong.

While Sydney and Brisbane had imperial and administrative origins, the other cities were founded by free settlers for commercial purposes. The Swan River Colony and the South Australian Company were established in 1829 and 1836 with the permission of the British government. Squatters from Van Diemen's Land (Tasmania) acted independently without government authority in establishing a settlement on the southern coast of the mainland, where Melbourne would be sited, in 1835. Perth was established on the land of the Noongar people, with an elongated grid of streets parallel to the Swan River to its south and a chain of lakes to its north. Limited water was available at the mouth of the Swan, and

[25] Cook, *A River with a City Problem*, 3–7.

FIGURE 3.1. The City of Melbourne, 1855. Nathaniel Whittock, artist (State Library of Victoria). Note the swampy ground south of the Yarra.

Britain's Colonial Office thought an upriver site would be easier to defend; a sand bar meant that a separate town, Fremantle, was required for ocean shipping. The town of Guildford was established at the furthest navigable point of the Swan, to serve as a port for rural settlement.[26] Colonel William Light's search for a site for the capital of the South Australia Company's venture found only one suitable harbour along the coast of South Australia, the present site of Port Adelaide, but it lacked fresh water. Light's town plan required a five-square mile site, and he chose slightly elevated ground on either side of a freshwater stream, Karrawirra Parri (River Torrens).[27] The river was a string of waterholes that drained into swampy reed beds near the coast and ceased flowing each summer. Melbourne's freshwater site was around 16 km from the Yarra's outlet at Hobson's Bay, where the port towns of Williamstown and Sandridge (Port Melbourne) would be located (see Figure 3.1). Fremantle and Port Adelaide were a similar distance from Perth and Adelaide.

[26] Bolton, 'Perth', 142–44.
[27] Williams, *The Making of the South Australian Landscape*, 398.

The Eora nation of Aboriginal peoples that lived in the coastal areas of Sydney had already been impacted by diseases carried by sealers, with half its population killed by smallpox in 1789.[28] In the same year, a drought caused Governor Phillip to order the excavation of three reservoirs in the sandstone banks of the creek at Sydney Cove; completed in 1802, these gave rise to the name Tank Stream. To protect the water supply, Phillip prohibited the felling of trees within 50 feet of Tank Stream and ordered the construction of a paling fence to keep out stock, but these regulations were difficult to enforce.[29] Urban demand for food and services provided economic opportunities for non-convict and ex-convict settlers. Shipping and trading activity shaped the town, as convicts, ex-convicts, and soldiers built huts or lived as lodgers in private households close to the shoreline and wharves. When Governor Macquarie arrived in 1809, Sydney was a helter-skelter settlement that had evolved organically with little evidence of civic responsibility.[30] Macquarie issued an order to demolish any business or drain that was fouling the Tank Stream, but already the settlement's waste was leaching into the groundwater.[31]

In *Replenishing the Earth*, James Belich argues that the Australian colonies and other 'Anglo-New Lands' were transformed by bursts of growth of the private settler population, which triggered cycles of rising economic activity to supply larger markets, followed by unstable expansion and boosterism, a slowing of growth as the boom collapsed, and then a 'rescue' through the growth of export-oriented industries.[32] These booms had a devastating impact on the Aboriginal population through disease, frontier violence, and the destruction of food sources. Recent population estimates for the entire continent show totals of 800,000 at first white contact, 500,000 by 1790, 300,000 by 1835, and 200,000 by 1850.[33] In the south-eastern corner of the continent, where river systems and food sustained an estimated 250,000 people, smallpox epidemics halved the population by 1790 and again by 1830.[34] The establishment of the squatter camp on the Yarra by land-hungry pastoralists from Van Diemen's Land in 1835 signified that 'the continent was fully opened to conquest'.[35] Occupation of land for pastoral use followed creek lines and

[28] Broome, *Aboriginal Australians*, 23; Butlin, *Economics and the Dreamtime*, 104–20.
[29] Broome, *Aboriginal Australians*, 248–50.
[30] Karskens, *The Colony*, 181.
[31] Cathcart, *The Water Dreamers*, 51–52.
[32] Belich, *Replenishing the Earth*.
[33] Hunter and Carmody, 'Estimating the Aboriginal population'.
[34] Butlin, *Our Original Aggression*, 146.
[35] Boyce, *1835*, xi.

natural water storages. A second boom began around 1847, as land sales and immigration to the Port Phillip District increased, with the gold rush accelerating the process. The Aboriginal population of the Port Phillip region was an estimated 60,000 in 1788; the official count in 1853 was fewer than 2000 – 'arguably the most rapid and catastrophic dispossession of the nineteenth century'.[36]

Having deprived Indigenous peoples of their best water sources and located fast-growing cities on their Country, the expansion of the pastoral industry, with its need for adequate water, meant the certain continuation of Indigenous dispossession. Settler Australians have only recently begun to acknowledge the dispossession of Indigenous people from their land, and that some of the most violent and destructive dispossession occurred over access to water. Many of the massacres that occurred over 150 years were near water sources, either because this is where Aboriginal people camped, or because access to the water was contested. Inland settlements, usually sited on river banks, dispossessed local Aboriginal tribes in a manner mirroring events that occurred in capital cities.[37] While for Aboriginal peoples (or nations), waterholes, streams, and sources were all part of the land and their identity, for white settlers, water was a critical asset to be consumed, controlled, and exploited. No better examples of this exist than in the five cities and the settler domestication of formerly Aboriginal water.

EMPOWERING SETTLER CITIZENS

The first hundred years of non-Indigenous settlement in Australia coincided with the Industrial Revolution and the era of the great industrial city in Western Europe and North America. In this relatively short time span, the world economy was transformed.[38] After the Napoleonic Wars, the leading European nations reduced trade barriers and forced open markets in Asia and Latin America. Steamships and railroads cut migration costs and freight rates; the telegraph provided faster commercial information. As economic growth stimulated demand for food and raw materials, surplus labour and capital generated by industrialisation was transferred

36 Broome, *Aboriginal Victorians*, 90–93; Smith et al., 'Fractional identities', 536.

37 Goodall, 'Riding the tide', 372–73; 'Colonial frontier massacres, Australia, 1780 to 1930', https://c21ch.newcastle.edu.au/colonialmassacres/map.php, accessed 7 June 2021.

38 Williamson, *Trade and Poverty*, 11–24.

by migration and investment to develop what Europeans and settler Americans considered to be new lands. During the nineteenth century, the value of world trade grew by 3.9 per cent per annum – three times the rate in the preceding century.[39] The Australian colonies, as part of the Western world's 'frontier', were built on the benefits triggered by these improvements to trade, transportation, and communication. Colonial citizens compared their situation to that of old-world cities and aspired to replicate the best ideas, networks, and standards found there, while attempting to avoid the worst aspects of crowding and poverty.

The transformation underway in the global economy was guided by the ideals of the Enlightenment: a set of beliefs and values that were positive (what the world is) and normative (how it ought to be), overturning old schools of thought and allowing new institutions to form. Adam Smith's view of individual self-interest – hitherto considered unseemly at best and sinful at worse – as a force that boosted the wealth of entire nations spawned movements against mercantilist restrictions on trade and industrial activity in Europe and the United States. David Ricardo's championing of free trade as the means of generating general well-being, rather than Corn Laws that benefited rent-seeking landowners, encouraged 'increasing resistance to any kind of measure that benefitted a few at the expense of the many'.[40] As ideas about the production of wealth emerged to fit changing circumstances, these in turn affected the ways in which the economic system mutated by establishing beliefs about social progress and prescriptions for the betterment of individual well-being that were impossible to undo.[41] New ideas about the role of the state in the function of the economy reflected the establishment of 'inclusive' institutions – secure property rights, effective legal systems, and investment in public infrastructure – that allowed people to make new business and career choices.[42] The emergence of Britain's market economy during the eighteenth century was built on pluralistic political institutions that distributed power through a broad coalition of interests, imposing constraints on the exercise of power.

For settler colonists, government provision of infrastructure, stable property rights, and relatively fluid class boundaries meant that no landed elite or political authority could capture all of the proceeds of resource

[39] O'Rourke and Williamson, 'Once more', 112.
[40] Greif and Mokyr, 'Cognitive rules', 40.
[41] Mokyr, *The Enlightened Economy*, 30–37.
[42] Acemoglu and Robinson, *Why Nations Fail*.

development. In European colonies where high mortality limited the number of settlers, as in Spain and Portugal's New World colonies, extractive institutions were likely to develop. Where settler mortality was lower and the Indigenous population was reduced by disease and violence, settlers created 'neo-Europes', with strong property rights, institutions, and active government involvement in the economic welfare of colonists.[43] Settlers wanted to build permanent homes and expected the same legal rights as existed in their home countries. These institutional frameworks evolved in a path-dependent way, exerting lasting influences on economic activity that sustained high settler incomes in colonies such as Australia and persistent poverty in other colonies.

Britain's governance of Empire was a dynamic process, with power exercised through bureaucratic structures and governors exercising discretionary powers working with the Colonial Office, rather than direct rule.[44] In Australia, the imperial government played an independent role in the evolution of democratic institutions as the convict share of the New South Wales population declined.[45] The British parliament created a New South Wales Legislative Council in 1823, beginning what became an orderly transfer of political power from the governor and his advisers to the colonists themselves. Once Parliament was satisfied that colonial subjects could benefit from greater political participation and operate elected institutions responsibly, the Australian colonies were permitted to progress to local self-government, with power to amend their own constitutions and the executive becoming ministers answerable to an elected parliament rather than the governor (Western Australia was the last colony to be granted self-government, in 1890). As property restrictions on franchise in lower houses were reduced, liberal democrats representing free immigrants, wage earners, and small business owners gained control of colonial governments.[46] South Australia introduced suffrage for Indigenous and settler women in 1894; after Federation, this extended throughout the Australian states and Commonwealth, but Indigenous women and men were largely excluded from the right to vote until 1962.

Participatory democracy gave voice to the demands of settlers, many with expectations of lives free of old world constraints, for improved

43 Acemoglu, Johnson, and Robinson, 'Colonial origins of comparative development'; Crosby, *Ecological Imperialism*; McNeill, *Mosquito Empires*.

44 Burroughs, 'Imperial institutions'.

45 McLean, *Why Australian Prospered*, 76–79.

46 McLean, *Why Australia Prospered*, 63–67; Hirst, *The Strange Birth*.

urban water infrastructure that benefited many more than a small elite.[47] Australia's GDP per worker grew rapidly until the 1890s, but the gains did not accrue disproportionately to squatters, landowners, and capitalists.[48] For most of the nineteenth century, a voyage from Europe to Australia took around three times as long and cost four times as that to New York; migrants to Australia were those with sufficient income to be able to afford high fares and live off savings while making the long journey.[49] Assisted passages that Australian governments provided from the 1860s to offset such high migration costs were generally not made available to illiterate and unskilled workers. As the cities grew, demand for unskilled labour in construction and loading and carting products for domestic use and external markets was high, and as migrants found work their expenditure on housing and other products added more to the level of aggregate demand than it did to the size of the workforce.[50] Growing urban markets created incentives for farmers to diversify and produce more efficiently, reducing the prices of a wide range of food products.[51]

Most immigrants to Australia came from British cities, seeking a quality of life that was unobtainable in their homeland.[52] From the mid-eighteenth century, London's merchant elite had moved from conducting business from city townhouses to new suburban villas that offered fresh air and privacy, where families could be shielded from crowded, dirty cities yet be close enough to commute to a city workplace.[53] As historian Graeme Davison observes, worker homes in nineteenth-century Australian cities 'were cut price bastard versions of the romantic ideal':

> Working-class immigrants seldom had the means to buy or build a grand villa; a crudely embellished timber cottage on an unmade street was often the best they could afford. But if the forms changed, the underlying expectations were often much the same. What the immigrants sought, and what their homelands often denied, was domestic independence. Freedom from the neighbours, freedom from the landlord, freedom from the boss: these, as much as the moral, aesthetic, sanitary and social ideals of the ideal English middle class, were the homeland dreams on which the Australian home was founded.[54]

[47] Brunkova and Shanahan, 'The Australian gold rushes'.
[48] Panza and Williamson, 'Australian squatters, convicts, and capitalists'.
[49] Davison, 'Colonial origins', 8. [50] Withers, 'Immigration and economic fluctuations'.
[51] Frost, 'The contribution of the urban sector', 55–56; Frost, *The New Urban Frontier*, 72–74.
[52] Davison, 'Colonial origins', 9. [53] Fishman, *Bourgeois Utopias*, 18–38.
[54] Davison, 'Colonial origins', 20.

This demand for independence, and the relative lack of poverty that helped make it possible, was initially conducive to private or local solutions to water management issues – such as wells and backyard cesspits. These solutions tended to pollute immediate neighbourhoods, encouraging citizens to seek larger scale public goods, such as networked water systems that relied on government support or direct provision. Australia's first colonial town councils lacked the resources to provide effective water infrastructure. In the first three decades of the New South Wales convict settlement, the British government maintained a high level of expenditure (40 per cent of gross national expenditure and 21 per cent of total employment in 1800–04) on public works, such as water supply, roads, and bridges, using convict labour. The share declined after 1800, but government expenditure still accounted for 17 per cent of spending and 8 per cent of jobs by 1830.[55] Transportation to New South Wales ended in 1840, and the economy soon entered recession. For the British government, the *Corporation Act* of 1842, establishing councils in Sydney and Melbourne, was an austerity measure, intended to reduce public spending by encouraging self-reliance in the provision of local amenities.[56] New South Wales, which governed territories extending over what became Victoria and Queensland, ignored requests from the Melbourne Corporation for funds to supply water to and drain the town and throughout the 1840s and 1850s neglected infrastructure in Brisbane.[57] The Adelaide Corporation, created in 1840, was £600 in debt by early 1842, unable to raise rate revenue from land that was purchased by speculators; Governor Grey sought to divest the Corporation of its power. The future of the colony was only secured after the British government wrote off the Corporation's debt, converting a loan to a gift, and as agricultural settlement spread beyond the Adelaide plains.[58]

Colonial legislatures allocated administrative tasks to city governments, but the process of investing in water infrastructure was slowed by a reluctance to transfer legal powers and resources to city corporations, which were seen as potential rivals. After gold was discovered in Victoria in 1851, Melbourne's population of 29,000 increased by 33 per cent each year in the subsequent decade, to 125,000.[59] The under-resourced

55 Boot, 'Government and the colonial economies', 99.

56 Fitzgerald, *Sydney 1842–1992*, 41–43.

57 Johnston and Gregory, 'Brisbane', 262.

58 Denholm, 'Adelaide', 187–88; Williams, *The Making of the South Australian Landscape*, 28–29.

59 McCarty, 'Australian capital cities', 119.

Melbourne Corporation made no attempt to annex areas of building activity beyond the city boundaries, and at the peak of the rushes in 1852–53, the Victorian government responded to housing shortages by exempting land outside the central grid from the *Melbourne Building Act* of 1849, which mandated fireproof construction and minimum street widths.[60] In Adelaide, high prices of city lots encouraged speculative buying and selling of land in around 30 villages within a four- to five-mile radius. Colonial governments in South Australia (1853) and Victoria (1854) decentralised local government by allowing the incorporation of municipalities that met requirements for area, population, and levels of civic engagement; the New South Wales *Municipalities Act* of 1858 was modelled on the Victorian legislation. Local artisans and small business owners formed committees, held public meetings, and petitioned colonial governments to improve roads and drainage and meet the requirement for the incorporation of new municipalities.[61] The Brisbane Town Council was a product of the New South Wales legislation, created shortly before separation in 1859.[62]

Inclusive governance structures and high overall living standards empowered citizens to appeal to governments for better environmental conditions. In Melbourne, the response to an 1863 flood that raised the Yarra 10 feet (3 metres) above its normal level, inundating mostly working-class areas adjacent to the river, reflected the capacity of citizens to articulate shared views and form relationships with local and colonial governments to address issues of flood relief and management.[63] Bottom-up action took the form of residents providing immediate assistance to neighbours and holding public meetings to establish relief funds and form distribution committees. These grassroots movements built relationships with local authorities and placed pressure on the colonial government to undertake flood mitigation works.

Residential segregation, embodied in a social divide between low-lying districts close to industrial areas and more affluent suburban retreats, created what environmental historian Dan Coward calls 'a queuing device' that determined access to urban infrastructure and services.[64] In municipalities where ratepayers could mobilise political power to protect land values and preserve local amenities, councils were able to resist the encroachment of manufacturing and noxious trades and the nearby siting

[60] Dunstan, *Governing the Metropolis*, 42–72.
[61] Barrett, *The Inner Suburbs*, 38–39.
[62] Laverty, *The Making of a Metropolis*, 94–97.
[63] Dobbie, Morgan, and Frost, 'Overcoming abundance'.
[64] Coward, *Out of Sight*, 55.

of waste disposal outlets, usually into waterways. These suburbs also benefited from access to public transport, schools, and hospitals. Inner suburban areas, where transport hubs, factories, warehouses, and worker housing were in close proximity, were impacted by pollution from open drains, sewer outlets, and gasometers. Poverty forced individuals to live in such areas, but the effects of poverty on conditions inside the home, such as household size, maternal age, and the resources and knowledge to practice hygiene and safe weaning, could make families vulnerable to infant death, even if low-density housing and safe water infrastructure outside the home were available.[65] As metropolitan areas expanded, the provision of network infrastructure became more expensive. The division of political authority between local councils across metropolitan areas would continue to shape government responses to water management challenges, with uncertain institutional arrangements and inequitable distribution of services and hazards presenting obstacles to improved amenities in settler cities.

CONTROLLING WATER

Sanitary reform, originating in Britain from the 1840s, was driven by fear of cholera in increasingly crowded cities.

> Cholera was swift, dramatic, highly lethal while it lasted, and extremely contagious: it struck terror into the minds of the middle and upper classes who ruled the cities and the country, and accordingly led, as no other did in the first half of the nineteenth century, to immediate, vigorous, administrative action.[66]

Across Europe, the public response to the first cholera epidemic in 1831–32 was to demand quarantining of affected areas. The sanitary reform movement, spearheaded by Edwin Chadwick's identification of the private market's failure to protect residents from disease, was based on the prevailing acceptance of miasma theory, which attributed diseases to vapours from rotting material and damp areas.

Sanitary reformers conceived the problem of disease in spatial terms, correctly identifying a correlation between disease, crowding, and poor sanitation, and advocating public investment in the construction of integrated sewerage and water systems to remove miasmas from the urban environment by channelling wastes out of the city and clearing of areas of

[65] Mein Smith and Frost, 'Suburbia and infant death'. [66] Flinn, 'Introduction', 10.

crowded housing.[67] After the second epidemic in 1848–49, Dr John Snow's 1853 experiment of removing the handle of a pump connected to sewage-contaminated water transformed popular attitudes to cholera from being a product of sin, poverty, and race, thus heralding a determination to make cities safe for their inhabitants.[68]

Sanitary reform had long been delayed by the resistance of industrialists, landlords, and workers themselves to the financial cost of reconfiguring urban landscapes. By 1848, British local authorities were empowered by the *Public Health Act* and other legislation to supply water and build sewerage systems, but this framework provided a set of options, not a mandate, that few towns had the resources to take advantage of.[69] Residents of low-income districts resisted communal solutions that were seen as likely to entail increased costs. Property owners, who stood to bear the financial burden of sanitary reform, lived in areas where the need for sanitary improvement was less urgent – water closets were widely used in affluent London suburbs as early as 1830.[70] Miasma theory provided helpful prescriptions for sanitary reform but was counterproductive when wastes were drained or dumped into waterways that provided drinking water. Water that looked clear and was not malodorous was considered safe, though it could carry disease. Chadwick and other sanitary reformers were aware of the importance of domestic management to health, through child care, nutrition, and household cleaning and hygiene, but considered the most pressing need to be in the public domain of streets, backyards, rivers, and sewers.[71] The emerging acceptance of germ theory shifted the emphasis of disease prevention from water-related public goods, such as drainage and safe water, to advocacy of measures such as improved ventilation, bathing and housecleaning, boiling drinking water (which tea drinkers did anyway), isolating sick members of the household, and the use of chemical disinfectants to address domestic sources of infection. Measures targeted at the poor, such as food inspections and housing and zoning regulations, caused the benefits of reform to trickle down the social ladder.[72] These ideas filtered through to Australia relatively quickly through individuals, families, newspapers, and official reports. While the Australian environment was not yet well understood

[67] Allen, *Cleansing the City*, 17–18. [68] Rosenberg, *The Cholera Years*, 5.
[69] Mokyr, *The Enlightened Economy*, 480.
[70] Baldwin, *Contagion and the State*, 539; Mokyr, *The Enlightened Economy*, 293.
[71] Mokyr, *The Enlightened Economy*, 293–96.
[72] Tomes, 'The private side of public health'; Baldwin, *Contagion and the State*, 539.

by the settler population, and the colonies' institutions, economies, and cities were far less developed than in Britain, they did have the advantage of not yet being burdened by decades of path-dependent urban development.

British municipal governments were able to secure finance for water infrastructure more easily after the 1860s, as the central government subsidised borrowing for town improvements and economic growth expanded local tax bases. Britain was free of cholera from 1866, and its lead was soon followed by North America, parts of northern and western Europe, and then the Australian colonies.[73] Despite increased commitments to reticulation, networked water supplies in British towns remained intermittent until the 1890s, and piped water was uncommon in working-class homes until at least 1914.[74] Only one in four American houses had running water in 1890, and supplies in many cities were drawn from waterways contaminated by sewage discharge. Filtration and chlorination of water did not begin in several US cities until after 1900.[75]

As newspaper accounts of cholera in Britain during the third pandemic reached Australia in 1849, anxiety over sanitary conditions in cities was heightened. In Melbourne, fear of cholera was driven by the town being 'without drainage of any kind' and so densely populated 'as to more resemble the state of things in an overpopulated country than a town which has sprung up in the wilderness within the last ten years'.[76] Over 400,000 people landed at Melbourne and Geelong during the gold rush between 1851 and 1860, and south of the Yarra, a cluster of tents and shanties known as Canvas Town housed homeless immigrants, including women with babies and young children left behind while men travelled to the diggings.[77] In 1851, the *Sydney Morning Herald* drew local parallels with conditions in British industrial cities, where the 'frightful disease, the cholera, like the advance guard of an enemy, has already warned us . . . no time should be lost in adopting temporary measures and coming to well-digested conclusions as to permanently putting the city into a good position in regard to its sanitary state'.[78] 'There is nothing to prevent the cholera from coming further eastward, and visiting Australia', an

[73] Hamlin, *Cholera*, 139–40.

[74] Smith, *The People's Health*, 228; Wohl, *Endangered Lives*, 62.

[75] Gordon, *The Rise and Fall of American Growth*, 122–25.

[76] Cited by Dunstan, *Governing the Metropolis*, 124.

[77] Frost, 'Water technology and the urban environment', 17; Davison, 'Gold-rush Melbourne', 60–61.

[78] 'The sanitary state of Sydney. No. X', *Sydney Morning Herald*, 5 April 1851.

FIGURE 3.2. River Torrens, looking south-west, c. 1860 (State Library of South Australia).

Adelaide reporter argued in 1853. 'It will be more to our interest to keep the cholera at a distance than to fight it when it comes'.[79]

The use of rivers as a source of drinking water was by now a pressing health concern. Adelaide drew water 'from the stagnant pools, or the muddy rapids of the Torrens – a great part of the year unfit for use, without filtration, and at all times expensive' (see Figure 3.2).[80] In Adelaide and Melbourne, private water carters drew supplies for weekly delivery to barrels near the front gates of houses, but the rivers were also used for bathing, watering horses and cattle, refuse disposal, and as an outlet for open drains and street channels carrying the contents of cess-pits.[81] Brisbane relied on tanks sunk in the government brickfields and a small dam across a creek at Tank Street, built using convict labour in 1828 and 1838, and wells dug by private residents. In Brisbane, Melbourne, and Adelaide, the quality of the water and its rising cost

79 'The sanitary state of Adelaide', *South Australian Register* (Adelaide), 28 October 1854.

80 'Supply of water', *South Australian* (Adelaide), 13 October 1848.

81 'Impurities of the Torrens', *South Australian Register* (Adelaide), 9 February 1854; Dingle and Doyle, *Yan Yean*.

was the subject of residents' letters of complaint to the press.[82] The Sydney Corporation extended reticulation from Busby's Bore, the system built by government order that connected water from Lachlan Swamp (now Centennial Park) by tunnel to a gravity-fed reservoir in Hyde Park, for distribution by water carts. Completed in 1837, Busby's Bore (named for the supervising contractor) was a stopgap solution, unable to meet the supply needs of the town. The Corporation's capacity to connect individual houses was limited by its borrowing power, which could not exceed average revenue over a five-year period.[83]

The Victorian government, in contrast, responded boldly, investing in water supply ahead of demand. A City Council review of all water supply and drainage methods in 1851 identified the Plenty River, a northern tributary of the Yarra, as the most appropriate water supply source.[84] The City Surveyor, James Blackburn, recommended a reservoir at Yan Yean, 29 km from Melbourne, which would 'render it needless to revert to the Yarra for any other purpose than as the principal drain of the country and sewer of the city'.[85] Blackburn estimated that the scheme could provide a constant supply to individual houses for 70,000 people – three times the current population.[86] In 1853, a Select Committee approved Blackburn's plan, with responsibility for sanitation transferred from the City Council to a Commission of Sewers and Water Supply.[87] The colonial government authorised a £600,000 loan to be repaid through rates levied on water users.

The supervising engineer, Matthew Jackson, amended Blackburn's plan by raising the height of the embankments to facilitate gravitational flows to 200,000 people. This exemplar of the British gravitation scheme of water supply was transplanted across the Empire; as urban historian John Broich has shown, 'at home and abroad, rulers and reformers identified the same practical problems ... ; in both places, British authorities saw reconfiguring the environment as a powerful tool for reforming society'.[88] When Yan Yean opened in 1857, beginning the city's

[82] Cossins, 'One hundred years', 2–3; Johnston and Gregory, 'Brisbane', 257; Cook, *A River with a City Problem*, 16.

[83] Clark, 'Worse than physic', 55; Fitzgerald, *Sydney 1842–1992*, 43–46; Coward, *Out of Sight*, 31.

[84] 'Report of the City Surveyor, No. 40', *Argus* (Melbourne), 26 May 1851.

[85] 'Water Supply', *Argus*, 26 May 1851. [86] Dingle and Doyle, *Yan Yean*, 7–15.

[87] 'Report from the Select Committee on the Sewerage of and Supply of Water for Melbourne', *Votes and Proceedings of the Legislative Council, 1852–53*.

[88] Broich, 'Engineering the Empire', 365.

dependence on its hinterland for its potable water supplies, Melbourne's population was around 100,000. Until the completion of the Lower Aswan Dam across the Nile (for irrigation use) in 1902, Yan Yean was the largest artificial lake in the world.[89] The system operated without threat of serious water shortage until 1875, by which time the metropolitan population was a quarter of a million.[90] At this time, Melbourne was the 60th largest city in the world.[91]

The priority of the Sydney Corporation was street improvement; water and sewerage planning were approached more cautiously. Engineers were scarce, and councillors wanted to avoid debt. The Corporation only laid water mains to wealthy residential areas and commercial and industrial users where initial revenue exceeded the capital cost. Most residents in the increasingly crowded city relied on fountains, pumps, wells, and water carters. From 1851 to 1856, the city's population rose from 44,000 to over 53,000 – 20.4 per cent, while the number of houses increased by only 11.9 per cent.[92] Only one in four households had piped water; even dilapidated housing near public water sources commanded high rents.[93] After a summer of water rationing in 1849–50, the colonial government in 1853 replaced the City Council with three City Commissioners. City manufacturers, who needed reliable water supplies to operate steam-powered machinery, petitioned the government to develop alternatives south of Sydney at Botany Swamp. An official inquiry considered proposals to draw water from the Nepean, Georges, or Hawkesbury Rivers, but these were either too expensive or had a long time lag. In recommending pumping water from the Botany wetlands, the inquiry underestimated Sydney's future population growth. Projected to reach 80,000 by 1871 by assuming the current rate of immigration would be maintained, the actual population reached 138,000. In this failure to account for the increased water needs of a sewerage system and water-using industries 'were sown the seeds of Sydney's water supply problem for the next three decades', economic historian David Clark observes. 'Once capital had been invested in the Botany project, the likelihood of work on other necessary but more far-sighted sources of supply was reduced'.[94]

89 Frost, 'Water technology and the urban environment', 21.
90 Dingle and Rasmussen, *Vital Connections*, 31–32; La Nauze, *Engineer to Marvellous Melbourne*, 69.
91 Chandler, *Four Thousand Years of Urban Growth*, 491.
92 Calculated from data in Coward, *Out of Sight*, 16.
93 Coward, *Out of Sight*, 14–17.
94 Clark, 'Worse than physic', 58.

When a reconstituted City Council took over from the City Commissioners in 1858, it inherited water supply debts that were difficult to repay, as legislation to allow the collection of water rates from all households was not enacted until 1879.[95] Water first flowed from Botany in 1859, but the pumps were not strong enough to reach elevated areas, and the Council was reluctant to extend mains beyond the city's boundary. Stormwater and dumped night soil polluted the Lachlan and Botany reserves. Effective control of the system rested with the engineer-in-charge, Edward Bell, who staked his reputation on the ability of Botany to cope with demand for many decades. As political pressure to find a larger, protected water source mounted, Bell refused to abandon the Botany scheme. As Coward observes, 'the City had spent £300,000 on the Botany works: it would, therefore, lose money if it had to construct works elsewhere'.[96]

After two years of drought and water restrictions, a Royal Commission into Sydney's water supply sat from 1867 to 1869. The Botany Swamp had only a small drainage area and was vulnerable in times of drought, while encroachment of industry and building activity undermined water quality. Bell increased the capacity of the Botany pumping station to deliver water to the Paddington reservoir in 1866 and built wooden dams on the Botany reserve in 1867. For the Commission, which estimated that Botany could supply a maximum of 250,000 people – just over half the current population – a storage on the upper Nepean River, transporting water 100 km to Sydney, was the superior option.[97] For ten years, construction was delayed until the issue of which level of government should build and control the waterworks was resolved. Parliamentary votes to transfer responsibility from the Council to colonial government were lost. The delay placed a heavier load on Botany, and when work on the Nepean scheme began in 1879, a race against time ensued to prevent Sydney from running out of water. Nepean water first reached Sydney in 1886, from temporary fluming that was required to avert a water famine.[98]

By the mid-1850s, both the Adelaide Corporation and colonial government were considering alternative sources of water. In 1855, the Corporation called for tenders for drainage of the 'dirty and neglected

[95] Clark, 'Worse than physic', 58. [96] Coward, *Out of Sight*, 82.
[97] Coward, *Out of Sight*, 84. Sydney's population would reach 225,000 by 1881.
[98] Clark, 'Worse than physic'.

city' and construction of a water supply for at least 30,000 people.[99] The sprawling city, with a population of only 18,000, had been 'laid out on a scale of unwise magnitude', a commentator noted. The 'unnecessary size of the town has augmented the cost and difficulty of providing an adequate water supply for the whole city'.[100] 'The absolutely best solution might not be very practicable', the *Adelaide Observer* noted:

> or, if practicable, only so at such a cost as would be prohibitory of it, and it may be wise therefore to be contented with something which, if short of perfection, is at least something within our reach, and which would be practically sufficient. ... The more simple and modest the plan we adopt, the quicker shall we enter into the enjoyment of whatever benefits it may confer.[101]

The Legislative Council appointed a Select Committee, which recommended a weir on the Onkaparinga River and sewer outlets in the Park Lands to drain the city, but the Corporation rejected this as too expensive.[102] The colonial government responded by removing the power to construct waterworks from the Council and transferring it to three commissioners, as in Sydney, and introduced a bill to borrow £280,000 for a reservoir on the Torrens, 11 km north-east of the city. The cost of servicing the loan through local rates was estimated at £18,000 per annum, well below the £28,000 citizens currently paid for impure water.[103] A Waterworks and Drainage Commission was appointed, with Victoria's Matthew Jackson invited to report on the proposed weir site.[104] After a flood destroyed the weir in 1858, the colonial government abolished the Commission and took direct control through the Department of Water Supply. A new reservoir was formed higher up the river, and the first supplies of reticulated water from Thorndon Park reached Adelaide homes in 1862.[105] Engineers were aware that the weir would have to be increased in size for water to supply mains throughout the metropolitan area.[106] A second reservoir on the Torrens at Hope

99 'Water supply', *Adelaide Times*, 26 February 1855; 'A move in the right direction', *Adelaide Times*, 26 January 1855.

100 Sinnett, *An Account of the Colony of South Australia*, 53.

101 'Supply of water to the City', *Adelaide Observer*, 19 January 1856.

102 'Water and sewage works', *South Australian Register* (Adelaide), 25 January 1856; 'The water supply of Adelaide', *Adelaide Observer*, 12 April 1856.

103 'Waterworks for the City', *Adelaide Times*, 11 June 1856.

104 'The drainage of Adelaide', *Adelaide Observer*, 26 December 1857.

105 Hammerton, *Water South Australia*, 5–16, 24–25.

106 'The reservoir at Thorndon Park', *South Australian Register* (Adelaide), 18 April 1859.

Valley was completed in 1872; service reservoirs were later built in the Park Lands to be filled at night to supply areas beyond the City.

After the creation of the Municipality of Brisbane in 1858, citizens called on the city to provide 'a constant and dependable supply'. One of the Council's first actions was to establish a Water Supply Committee, 'to inquire into the best means of procuring a permanent supply of water to the town, including both sides of the river'.[107] The Committee commissioned a report by Thomas Oldham, a British civil engineer who had worked in Victoria, who recommended a dam across the river above Brisbane, capable of supplying 20 gallons (91 litres) per day to 150,000 people (at the time, Brisbane's population was around 6000). The Queensland government had insufficient funds for works of that scale and, in 1863, created a statutory authority, the Brisbane Board of Waterworks, removing responsibility for water supply from the Council. The Board drew up plans for an earth dam on Enoggera Creek, linked to reservoirs closer to the city, to supply a population of 25,000. To reduce expenditure further, the Board modified its original plan to reduce the trunk main from 12 to 8 inches in diameter, limiting the population that could be supplied to 15,000. Enoggera Dam was completed in 1866, but plans to supply South Brisbane were abandoned after a flood the following year.

Brisbane was in the grip of drought for much of the decade from 1880 to 1889. When reticulation extensions were suspended as the system provided inadequate supply, a failure to keep pace with the water needs of a growing city was apparent. One in four of an estimated population of 70,000 in 1888 had no access to reticulated water. The Board responded to this lag and the continuing drought by investing ahead of demand. Gold Creek Dam, completed in 1886, doubled the capacity of the system; in 1888, the Board built a further reservoir at Mount Crosby, with a pumping station to draw water from the Brisbane River. The scheme was intended to supply 300,000 people with 40 gallons (182 litres) per day.[108] Water quality, however, remained poor, and most households continued to use rainwater tanks.[109]

In Perth, the provision of reticulated water was delayed by the colonial government's lack of resources and the view, as in England, that the issue

[107] Cited by Cossins, 'One hundred years', 3. See also Cook, *A River with a City Problem*, 16–18; Laverty, *The Making of a Metropolis*, 60–61.

[108] Cossins, 'One hundred years', 5–10.

[109] Lawson, *Brisbane in the 1890s*, 12.

was a municipal responsibility.[110] An 1885 Sanitation Commission recommended that the government prepare, and discuss with private contractors, a scheme to supply water from the Darling Ranges, to be built when funds were available. Outside the City of Perth, municipalities with local supplies were opposed to such investment. Fremantle Council laid pipes from wells and tunnels hewn into the limestone to access groundwater under Fremantle Prison. Residents in more affluent Cottesloe and Claremont had access to water from a private bore from the late 1890s. In 1889, ratepayers in the City of Perth voted to call for submissions for a water supply scheme. The Council reached agreement with a Melbourne-based syndicate, which would build a reservoir in the Darling Ranges, leading to a city service reservoir at Mt Eliza. Victoria Dam was completed in 1890, but the pipe connecting to the city reservoir was too small to provide adequate supplies, and the catchment was polluted by cattle grazing and timber milling. This mix of private and local government infrastructure provision was ineffective, and with the quality of water in the metropolitan area deteriorating, in 1896 the government took over the scheme, establishing a Metropolitan Waterworks Board at a cost of £220,000. Supplies were still inadequate, and in 1903, Parliament voted to take full control. Many households relied on tank water, but after the State government had reinvented the Board as the Metropolitan Water Supply, Sewerage, and Drainage Department (MWSSDD) in 1910, work began on the construction of new service reservoirs, increasing the capacity of Victoria Dam, sinking bores, and reticulating new suburbs.[111]

DIRTY WATER

'For the government of a state', historian Christopher Hamlin observes, 'the most important thing to do in the face of cholera was to *be seen* to be doing something, to be implementing some policy in which hope could be invested. … The precise action taken mattered less and its success almost not at all.'[112] Policies that were more gesture than well-planned action were likely to have unintended negative consequences. The New South Wales *Sewerage Act* (1850) exemplifies this point. The Act empowered the Sydney City Council to build sewers, but borrowing was limited to

[110] Hunt, *Water*, 20–40.
[111] Hunt and Bolton, 'Cleansing the dunghill', 14; Hunt, *Perth's Early Water Supplies*, 11.
[112] Hamlin, *Cholera*, 105. Emphasis as in original.

£20,000 and only households that agreed to connections were required to pay. 'Goading the City towards a sewerage policy was symbolic: initiative could be claimed but in reality little would be achieved.'[113] By the mid-1860s, the City had built five main brick sewers, which resulted in the covering of natural creeks, to drain water closets and storm water to the Harbour. An 1866 Royal Commission on harbour pollution found that the lack of strong tides and accumulation of silt turned shallow areas around the outlets into 'immense cesspools'.[114] The Commission concluded that sewer outlets away from the Harbour were required, but this was beyond the capacity of the indebted Council. Colonial government intervention would be required.

The Yarra's deterioration from a river to sewer epitomises the issue of the limited powers and resources available to the agencies charged with providing water and sanitation. Melbourne's Commissioners of Sewers and Water Supply followed Sydney's example in constructing surface and underground drains to remove storm water along lines of natural drainage to the nearest waterways. Private household wastes were deposited in cesspits. City surveyor James Blackburn told an 1853 Select Committee that if Melbourne's population continued to be 'moderately numerous' – around 100,000 – 'there would be no objection to using the Yarra as the main sewer. This would very much simplify the scheme of Sewerage'.[115] This was standard practice in British and American cities, but population growth and the connection of water closets to sewers made urban environments prone to local flooding and pollution.[116] As Blackburn warned, 'if Melbourne should become a very large city' – he suggested one with a population of half a million – 'the Yarra would be a nuisance'.[117] The colonial government vested all Crown lands in a Board of Land and Works, with statutory responsibility for the construction and maintenance of public buildings, roads, railways, bridges, water supply, sewers, and drains. The Board did not take direct responsibility for all of these activities and distributed grants to municipal councils, which did not have borrowing powers, for road building and drainage.

London was the city the Australian colonies sought to emulate. London's role as the British Empire's leading city raised aspirations in the Australian colonies. The creation of the London Metropolitan Board

113 Coward, *Out of Sight*, 33–34. 114 Coward, *Out of Sight*, 97.
115 'Report from the Select Committee', 21.
116 Tarr, 'The separate vs. combined sewer problem'.
117 'Report from the Select Committee', 21.

of Works in 1855 to coordinate public works across municipal boundaries and the construction of engineer Joseph Bazalgette's underground sewerage system in 1865 confirmed water-carriage as 'the standard to which most Australian sanitary experts aspired', as Davison puts it. 'Australians, like Britons and Americans, would come to regard the water closet and the flush cistern as indispensable markers of civilisation'.[118] In Sydney, the relative merits of water-carriage and dry conservancy (earth closet) systems were contested.[119] Advocates of earth closets opposed water-carriage not only because of its likely effect on water pollution but also because it risked wasting a valuable resource. Earth closets had greater potential for reclaiming fertiliser but were labour-intensive, requiring manual loading of soil and emptying of contents. Several patent earth closets were produced and offered with maintenance packages; one was trialled successfully at Brisbane Town Hall as late as 1906.[120] Yet, most government officials, engineers, and, ultimately, the public wanted capital-intensive sewerage that worked automatically without the need for manual work or monitoring. A writer in the *Brisbane Courier* contended in 1900 that dry-earth closets were 'totally unsuited to a city with the pretensions of Brisbane. An up-to-date water-carriage system of sewerage is imperative for the capital, and the lack of it is a subject for much comment of visitors from the Southern capitals or from abroad'.[121]

The major obstacles to the building of sewerage systems were the costs of extending infrastructure across dispersed metropolitan areas and the fragmentation of political authority between local councils. In Adelaide, where planning for a reticulated water supply in the mid-1850s included the 'draining' of the town, the issue was complicated by the lack of a convenient outlet. 'The Torrens is not an ordinary river, running with a full body of water to the sea', observed a writer in the *South Australian Register*, 'but a stream that spends itself on low-lying ground between the city and the beach'.[122] A Select Committee proposed that sewers would lead to tanks in the Park Lands, where solid waste would be separated

118 Davison, 'Down the gurgler', 40–41.

119 Beder, 'Early environmentalists and the battle against sewers in Sydney'. Prior to the introduction of sewerage in Melbourne, one proposed solution was the use of desiccator machines to reduce nightsoil to a powder, for sale as fertiliser, but this failed due to a lack of agreement between municipal councils over where plants would be located. Taylor, 'The "desiccator difficulty"'.

120 'Sanitary improvement', *Telegraph* (Brisbane), 6 June 1906.

121 Cited by Lawson, *Brisbane in the 1890s*, 14.

122 'Impurities of the Torrens', *South Australian Register* (Adelaide), 9 February 1854.

and deodorised for sale as manure, with filtered liquid draining to the Torrens, at an estimated cost of £175,000. A proposal by the Waterworks and Drainage Commissioners for disposal at sea was costed at £250,000. Authorised to borrow only £280,000, the colonial government abandoned both drainage proposals, directing the entire loan to water supply.[123] For the *South Australian Register*, a drainage system could safely be delayed. 'To suit our immediate wants, we require some remedy more promptly applicable'.[124] In Melbourne and Sydney, councils representing inner suburbs resisted increasing taxes that might discourage local industry. Outer suburbs did not perceive a need for a sewerage system, because miasma theory promised that elevation and sea breezes, and relatively large lots with space for backyard cesspits, would provide protection from disease.

The first indication of a more collaborative governance approach to sanitation problems was the New South Wales Premier's appointment of a Sewage and Health Board in 1875, in response to an epidemic of scarlet fever and measles in Sydney. The Board comprised the commissioners who had recommended the Nepean scheme in 1869, health officers and medical practitioners, and representatives of the City and municipal councils. There was now common ground between colonial and local government over the complementary nature of sewerage and water supply, recognition that the sewerage system would need to cross municipal boundaries, and that the colonial government was critical to financing and constructing the system.[125] The Board proposed a two-part solution: a trunk interceptor diverting sewage from the Harbour to the Pacific at Bondi and a combined sewer for southern-sloping municipalities to a sewage farm at Botany Bay. The recommendation for a combined Botany sewer reflected the area's smaller volume of waste and the absence of existing storm water drains; the sewage farm reflected a desire to experiment with land-based methods of disposal with output used as fertiliser. The Board considered the Nepean water supply scheme to be essential to the effective flushing of the sewers. The colonial government appointed William Clark, a British engineer who had designed a sewerage system for Calcutta, to assess the water supply and sewerage proposals.[126] Clark spent over a year in Australia, arriving in Sydney in late

123 'The drainage of Adelaide', *Adelaide Observer*, 26 December 1857; 'The City drainage', *South Australian Register* (Adelaide), 14 August 1856.

124 'The sanitary state of Adelaide'.

125 Coward, *Out of Sight*, 101.

126 Anon., 'Obituary. William Clark, 1821–1880'.

1876 and early the following year travelling to Adelaide to report on a proposal to connect the city sewers, which drained to large cesspools in the Park Lands, to a sewage farm.[127]

Clark's reports made a strong case for abandoning labour-intensive collection methods on financial grounds. Clark calculated the total cost of emptying and cleaning Sydney's 9924 cesspits (in 59 per cent of the housing stock) twice a year as equal to interest payable (at 5 per cent) on the loan required to build the Board's sewerage system.[128] In Adelaide, where Clark estimated the annual total cost of cesspit cleaning at £11,099 and interest on the cost of drainage work at £7250, work on the sewerage system began in 1878.[129] Construction of Sydney's southern main and Bondi outlet by the Public Works Department began in 1882, and when completed in 1888, ownership of the system, together with the Nepean waterworks, was transferred to a Metropolitan Board of Water Supply and Sewerage. In 1889, the colonial government approved a western suburbs sewerage system, designed by George Stayton, a Public Works Department engineer. Extending from the western suburban fringe, then south to an enlarged Botany sewage farm, Stayton's network of sewers was built ahead of demand, draining an area three times as large as that of the Bondi and Botany main sewers.[130] Assuming the existing rate of population growth of the western suburbs would be maintained, Stayton projected optimistically that the system would reach maximum population capacity around 1907; by 1933, that threshold had still not been reached. The Board met the extra costs through higher uniform rate charges across the entire system, effectively transferring resources from poorer, inner-city ratepayers to more affluent suburban ones.[131]

Melbourne added over 200,000 inhabitants during a spectacular economic and real estate boom, almost doubling its population between 1881 and 1891. Seventy per cent of this population growth was absorbed by new, largely middle-class suburbs built on sloping and seaside areas to Melbourne's south and east and flatter, working-class suburbs to the north and west.[132] Despite a safe water supply, inadequate sanitation kept the city dirty. Typhoid death rates fell below 60 per hundred thousand people in only four years from 1871 to 1890 and reached 120 during

[127] W. Clark, 'The drainage of Adelaide', *South Australian Register* (Adelaide), 12 January 1878.

[128] 'Report of Mr. W. Clark on the drainage of Sydney', *Sydney Morning Herald*, 20 July 1877.

[129] Clark, 'Drainage of Adelaide'.

[130] Coward, *Out of Sight*, 128–31.

[131] Coward, *Out of Sight*, 135.

[132] Davison, *Marvellous Melbourne*, 183–84.

the epidemics of 1879 and 1889.[133] A Royal Commission recommended that a statutory body, the Melbourne and Metropolitan Board of Works (MMBW) comprising representatives of all municipal councils, be created to build a sewerage system and operate the city's water supply.[134]

Melbourne's typhoid death rate was falling before the first houses were connected to the sewerage system, which suggests that households, newly armed with information about germs and the spread of disease, modified their behaviour through actions such as boiling water, washing hands before preparing or consuming food, and using flywire screens to prevent contamination of milk and household utensils.[135] Changing household expectations drove popular support for networked infrastructure, not rising costs of manual waste disposal. Economic historian David Merrett calculates that the anticipated cost of sewerage to Melbourne households was double that of pan collection.[136] Ratepayers were required to have connecting drains built at their own expense, and these up-front charges, together with the cost of lavatories and cisterns, would have met the costs of pan collection for another decade. As in Sydney and Adelaide, where Clark had underestimated the cost of sewerage by not accounting for individual household connection expenses, falling costs of manual cleaning weakened the case for capital-intensive sewerage. In Sydney, uniform sewerage rates imposed a burden on low-income household budgets that was used to subsidise the provision of water infrastructure in middle-class suburbs.[137] In Melbourne, the first suburbs to be sewered paid higher MMBW rates than outer suburban properties without sewerage.[138] Water and sewerage in Adelaide was a combined operation under the control of Engineering and Water Supply Department:

> As fast as the new houses spread in the suburban blocks the public authorities provided essential facilities which later included gas and electricity; in a way, there was an undeclared partnership between the individual private housebuilder and the larger public authority to propagate the new environment of the suburbs.[139]

The introduction of water-carriage sewerage was often discussed in Brisbane in the second half of the nineteenth century but 'seemed a

133 Frost, 'Water technology and the urban environment', 25.
134 Dingle and Rasmussen, *Vital Connections*; La Nauze, *Engineer to Marvellous Melbourne*, 127–36.
135 Frost, 'Water technology and the urban environment', 26.
136 Merrett, 'Economic growth and well-being'.
137 Coward, *Out of Sight*, 121.
138 Dingle and Rasmussen, *Vital Connections*, 66–69.
139 Williams, *The Making of the South Australian Landscape*, 434.

visionary proposal because of the lack of finance for civic purposes'.[140] The *Health Act* of 1872 required local boards of health to empty and if necessary close cesspits throughout the city. As in Perth, earth closets were seen as the next best alternative.[141] Throughout the 1870s, the Perth and Fremantle Councils' Inspector of Nuisances recommended dry-earth closets, regular collection by contractors, and disposal at a sewage farm. Perth Council acted to encourage installation of earth closets but chose not to increase general rates to meet extra costs of sewage collection.[142] The 1885 Sanitation Commission opposed the construction of sewers due to the lack of strong tides in the Swan River. As in Brisbane, councils introduced pan collection services by the early 1890s, but houses were still able to use cesspits and there were complaints of slovenly practices by night-men, such as washing pans in waterways. Perth's typhoid epidemic during the colony's gold rush, and the wave of immigrants from Victoria who expected Melbourne standards of water infrastructure, resulted in an 1897 inquiry into potential sewerage schemes. The estimated cost was £500,000, and, having invested five times that amount to supply water to the Kalgoorlie goldfield, the colonial government was unwilling to incur further debt.[143] Experiments with septic tanks demonstrated the feasibility of bacterial systems, and in 1904 the State government began planning for a sewerage system for a population of 30,000, with a Treatment Works on Burswood Island in the Swan River near Claisebrook.[144]

ENTERING THE TWENTIETH CENTURY

By 1901, more than 1 million people living in Sydney, Melbourne, Adelaide, Brisbane, and Perth – over 75 per cent of the five cities' population – had domesticated water supplied to their homes. The network's distribution, however, was far from complete. While almost two-thirds of Sydney's residents had constant piped water, in Perth the proportion was only around one-quarter. The average daily quantity of water used by the almost half a million people in Melbourne receiving piped water was around 265 litres per person and in Perth 205 litres.[145] Sewerage

140 Laverty, *The Making of a Metropolis*, 60.
141 Laverty, *The Making of a Metropolis*, 149–50.
142 Hunt, *Water*, 15–17.
143 Kelly, 'The development of housing', 54, 189; Hunt and Bolton, 'Cleansing the dunghill', 13.
144 MWSSDD, *Annual Report 1906*.
145 Calculated from *Official Yearbook of the Commonwealth of Australia* (various issues), Local Government and Water Supply and Sewerage. Figures are exclusive of Adelaide, where data on meterage was not available until 1919.

networks were less extensive and varied between the cities. By 1911, Adelaide was close to fully sewered, as were Melbourne (87 per cent of households) and Sydney (83 per cent). Only 700 properties were connected to sewerage in Perth, and none in Brisbane.[146] Perth's experiment with private and fragmented local provision of water was over – responsibility then passed from municipalities to a metropolitan board, as in Sydney, Melbourne, and Brisbane. In Adelaide, systems were managed by a Commissioner of Public Works. Most of these networks had been in place for less than 30 years. True domestication of urban water in Australia had seen several hesitant starts and was still a work in progress.

Cleaner streets, yards, and waterways were part of an overall gain in living standards.[147] In the mid-nineteenth century, when the five cities' populations were smaller and residential densities lower, people had obtained drinking water privately, from local waterways and wells, disposing of waste in their own cesspit. While these initial solutions preserved the independence of households and were cheap and flexible because start-up costs were low, toting water and waste was also a disincentive to cleanliness, polluting the natural and built environment. Sanitary reform originating in Britain in the 1840s highlighted problematic interconnections between water and waste, making the case for large-scale planning of public water and sewerage systems to prevent disease and poverty.

In most nineteenth-century cities, fear of cholera and later typhoid shaped the focus and spending of governments and the expectations of citizens and fostered the conditions for reshaping the relationship between the individual and the state. Once connected to the municipal water supply, the domestic sphere was drawn into a wider technological web that facilitated the extension of liberal reform into the private home. This bridging of the domestic and public spheres now enabled the liberal governance of the individual, such that 'sanitation was always a "social question"', as social historian Patrick Joyce argues for nineteenth-century Manchester and London. 'The case of the city and the care of body became as one, just as the health of the city and health of the body were one'.[148] By directing wastewater within the home, the sanitary infrastructure of the city cemented the embrace of the public and private.

Sanitary solutions both reflected and shaped spatial changes to cities. Damming rivers and accessing piped water and draining waste from city centres to low-lying areas or to waterways advantaged some areas more

[146] Neutze, *Urban Development in Australia*, 38. [147] Shanahan, 'Wealth and welfare'.
[148] Joyce, *The Rule of Freedom*, 65.

than others. While living standards improved overall, residents with means, living closer to the networks, did better than others. Reform was advanced, or delayed, by politically determined trade-offs between private costs and the public provision of networks financed though rates or charges. Rising population density increased the total cost of private provision. Labour-intensive methods became more expensive as water and waste needed to be moved further, at greater external cost, transmission of disease was faster, and the numbers affected were higher. More people also created economies of scale, lowering the per capita cost of capital-intensive infrastructure. Rudimentary systems whose initial appeal was a lower price transferred costs to later generations. As in Britain, pan toilets were an expedient – a more effective solution to sanitation than cesspits and open drains, cheaper than water-carriage sewerage, and with the possibility of use as fertiliser.[149] As population and waste increased, sewerage became inevitable – but at much higher cost, especially in low-density cities and where there were no tidal rivers or large bodies of seawater for convenient disposal. Combined sewers were initially cheaper than separate systems but costly later, both financially and because of problems caused to nearby waterways and water supply intakes.

When Australia's first major public reservoir – Melbourne's Yan Yean – was approved in 1853, many of the conditions that would shape future investment in all capital cities were already in place. Water sources that had determined the cities' locations were already proving inadequate. Rapid growth was exposing the problems caused by private waste disposal. Each city faced these issues with constraints that included their particular topography, institutions, and capacity to cooperate and fund chosen solutions. Attitudes to sanitation and aspirations of the citizenry were powerful bottom-up forces that reflected inclusive, liberal democratic institutions and high overall living standards. In each city, the efficacy of solutions to common resource problems through the provision of water infrastructure was the outcome of cumulative decisions made over time. Such path-dependent networks not only reflected earlier solutions but also constrained the options available to future decision makers.

[149] Daunton, *House and Home in the Victorian City*, 252–60.

4

Keeping Up

When William Thwaites died of pneumonia in 1907 at age 53, construction of his planned sewerage system for Melbourne was almost complete. Born and trained in Melbourne, Thwaites was only 37 when he was appointed as the first engineer-in-chief of Melbourne and Metropolitan Board of Works (MMBW), 'very much the local boy who had risen to the top in a hurry'.[1] 'The fingerprints of William Thwaites can be found in every aspect of the engineering work', biographer Robert La Nauze writes. 'His signature appears on almost every plan'.[2] Thwaites retained personal control over work on the pumping station, sewage farm, and sewers, using the new technology of the telephone to call foremen at construction sites each morning to check on progress, as well as overseeing the engagement of contractors and procurement of materials. The use of local firms, often on small sequential contracts, placed Thwaites 'in the position of conductor of the orchestra, a role he relished'.[3] 'To those acquainted with his habits', a writer in the *Argus* observed, 'it is well known that he shortened his life by incessant labour in the service of the public'.[4]

As engineer to Melbourne's Water Supply Department, Thwaites and two other Melbourne engineers presented plans to the 1889 Royal Commission into Melbourne's sanitary condition for a system to convey sewage to a farm at Werribee. All three plans proposed systems with main

[1] Dingle and Rasmussen, *Vital Connections*, 49.
[2] La Nauze, *Engineer to Marvellous Melbourne*, 144.
[3] Dingle and Rasmussen, *Vital Connections*, 71.
[4] 'William Thwaites', *Argus* (Melbourne), 26 November 1907.

sewers for sewage only, running along the lines of creeks and rivers, with the sewage farm necessitated by the lack of strong tidal action in the Yarra and Port Phillip Bay. The Royal Commission appointed a British engineer, James Mansergh, to consider these proposals and draw up plans for a sewerage system. Mansergh's plan, probably influenced by conversations with Thwaites and data supplied by him, was designed to meet the needs of a population of 1.7 million without the need for extension of the main works – a figure Melbourne was projected to reach by 1939 (based on the assumption that Melbourne's current high population growth would continue, the estimate was optimistic – the city did not reach that mark until before the 1961 census).[5] Mansergh favoured the creation of two sewerage districts, with areas west of Gardiners Creek (a southern tributary of the Yarra) connected to Werribee, and eastern and southern suburbs to a sewage farm at Mordialloc. As loan finance became more difficult to obtain on the London market during the depression of the 1890s, Thwaites resisted pressure to sewer only the inner suburbs and modified Mansergh's scheme by reducing the capacity to 1 million people, connecting all mains to a pumping station at Spotswood, then to the Werribee farm.[6]

Work proceeded slowly, with the first connections made in 1897. To maintain quality, Thwaites issued bylaws for plumbers and negotiated contracts with hardware suppliers, placing standardised orders for all fittings, from earthenware pipes to complete weatherboard buildings for external water closets. Thwaites declared an aim to connect 15,000 houses each year, but this was overly optimistic given statutory limits on the Board's borrowing and rating capacity. In only one of the subsequent thirteen years was Thwaite's target close to being met. Nevertheless, in nine years between 1900–01 and 1909–10, a total of 67,000 houses were connected, increasing the proportion of the metropolitan total with sewerage from 42 to 86 per cent.[7] After Thwaites' death, a new chairman and engineer-in-chief 'assumed that the Board's primary task of building the sewers was complete; what was now required was frugal management'.[8] But population growth was creating increased demand for housing and related infrastructure. 'The race to water and

[5] Dingle and Rasmussen, *Vital Connections*, 42.

[6] Despite this reduction in planned capacity, the system continued to operate effectively until the 1960s, when Melbourne's population reached 1.8 million. La Nauze, *Engineer to Marvellous Melbourne*, 144.

[7] Dingle and Rasmussen, *Vital Connections*, 102–03.

[8] Dingle and Rasmussen, *Vital Connections*, 109.

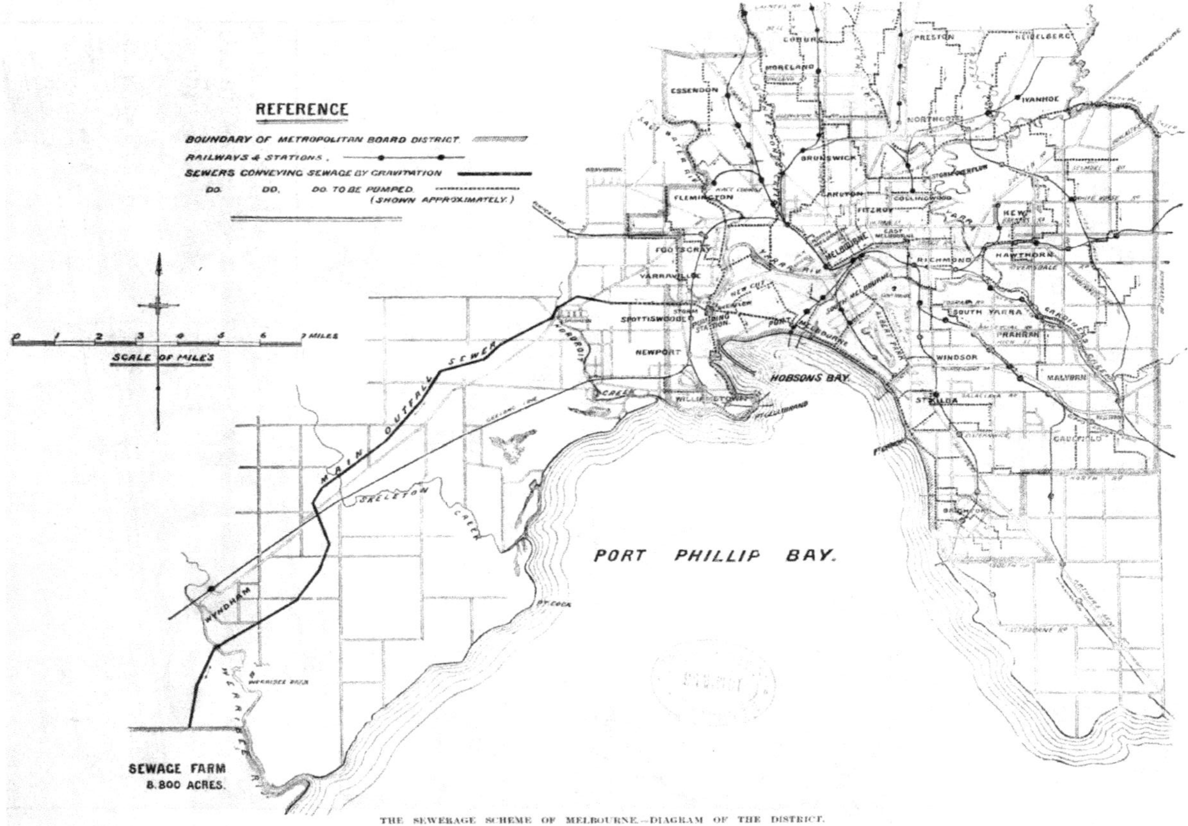

FIGURE 4.1. Sewerage scheme of Melbourne, 1892 (State Library of Victoria).

sewer a growing city was on', observe Tony Dingle and Carolyn Rasmussen in their history of the MMBW. 'It was a race which the Board would run whenever Victoria was prosperous during the next eighty years.'[9]

In this chapter, we consider the provision of water infrastructure in the first half of the twentieth century, in particular the ability of water bureaucracies such as the MMBW to keep up with rising demand caused by population growth and the expanding suburban footprint of the five cities. Water infrastructure that serves entire metropolitan areas constitutes a natural monopoly – large capital requirements and the indivisibility of investment make it cheaper for one provider to supply the market than to construct two or more duplicate services. In Australian cities, government authorities had the advantage of usually owning most of the land that water pipes and sewer mains had to pass through, while also having the power to compulsorily acquire land for such purposes if needed. Water supply and the disposal of waste make use of natural resources that are subject to diminishing returns. That is, without technological progress in water storage, delivery, and disposal, urbanisation creates rising costs for networks as local supplies and outlets are eventually exhausted.[10]

For a private company, maintaining profits would depend on an ability to increase charges or to capture a larger number of consumers. For a natural monopoly, diminishing returns can be offset by economies of scale, which reduce total costs as the scale of production increases. The limited range of substitutes for water services creates the potential for natural monopolies to sustain profits over long periods. Economist Max Neutze observes that there is a strong economic case for collective provision of water infrastructure but also for individual consumers to pay a price related to the cost of provision, given the private benefits that result from clean water and safe sewage disposal. 'In these circumstances, it is desirable that, as far as possible, hydraulic services should be sold to users at an appropriate price rather than provided free and funded from tax revenue.'[11]

Water and waste disposal services in Australia have never been provided to households as pure public goods, with no charge to users (as distinct from services in public places, such as drinking fountains

[9] Dingle and Rasmussen, *Vital Connections*, 103.
[10] Millward, 'The political economy of urban utilities', 331.
[11] Neutze, 'The provision of hydraulic services', 64.

and public toilets). Australia's water and sewerage authorities levied rates on all properties, establishing captive markets that were dependent on public supplies for potable water. Sewerage authorities and local governments used several mechanisms to establish control over users of sewerage services: setting charges based on property values, wholly or partly disregarding the volume of consumption; requiring households to be connected; regulating the use of alternate services; and discriminatory charging systems that could be revised after connection. Given the impracticality of collecting revenue from consumers (the free rider problem), most water infrastructure was financed by public borrowing, with loans amortised from rate revenue by averaging the costs of service over all properties in supply areas. Household decisions about water consumption and sewage load were price inelastic – with changes in quantities largely unresponsive to changes in rates and charges – which allowed authorities to increase revenue when rising costs needed to be covered.[12] This funding model had some equity advantages: annual charges based on property values allow households to spread costs over the course of a family lifecycle. But such charges historically provided little incentive for consumers to economise on water use or reduce the volume of sewage discharged and little incentive for utilities to discourage use and failed to account for the environmental costs of resource use, such as reduced river flow and damage to water quality from the discharge of effluent.[13]

Because water demand is to a large extent habitual – everyday patterns of washing, waste disposal, and garden watering are path-dependent, drawing on established cultures, infrastructure, and technology – policies to encourage a reduction in demand have had a limited impact in the short-term. As a result, Australian water utilities tended to react to water shortages by imposing restrictions on outdoor water use, usually followed by longer-term investment in major infrastructure to increase the amount and security of water supply.[14] Each of the five cities faced water shortages in the early 1920s. Twelve major water storages to serve metropolitan areas were built, or their construction commenced, during the interwar period (see Table 4.1).

The autonomy of water authorities varied between cities and over time.[15] Dominated by rural members who were indifferent, if not hostile,

[12] Butlin, Barnard, and Pincus, *Government and Capitalism*, 244.
[13] Neutze, 'The provision of hydraulic services', 64.
[14] Leroux, Martin, and Zheng, 'Addressing water shortages by force of habit'.
[15] Butlin, Barnard, and Pincus, *Government and Capitalism*, 245.

TABLE 4.1. *Major water storages completed, or construction commenced, 1921–47*

Storage	Dates	City	Authority
Cordeaux Dam	1918–26	Sydney	MWSDB
Avon Dam	1921–27	Sydney	MWSDB
Maroondah Dam	1920–27	Melbourne	MMBW
O'Shannassy Reservoir	1922–28	Melbourne	MMBW
Churchman Brook Dam	1923–29	Perth	Metropolitan Water Supply, Sewerage and Drainage Department (Western Australia)
Nepean Dam	1925–35	Sydney	MWSDB
Warragamba Weir	1925–40	Sydney	MWSDB
Silvan Reservoir	1926–31	Melbourne	MMBW
Woronora Dam	1927–41	Sydney	MWSDB
Mount Bold Reservoir	1932–38	Adelaide	Engineering and Water Supply Department (South Australia)
Canning Dam	1933–40	Perth	Metropolitan Water Supply, Sewerage and Drainage Department (Western Australia)
Somerset Dam	1935–53	Brisbane	Bureau of Industry, Stanley River Works Board

to Sydney and its infrastructure needs, the New South Wales parliament limited Metropolitan Water and Sewerage Board powers over bylaws, catchment areas, and finance. Expenditure was restricted to small reservoirs, pumping stations, and mains, required the approval of the Minister of Public Works, and was restrained by an annual parliamentary vote; all rate revenue was directed to consolidated revenue. The Public Works Department retained control of all major works and constructed all water infrastructure. Public dissatisfaction with the reliability of water supplies prompted the removal of these restrictions in 1924, with a reconstituted Metropolitan Water and Sewerage and Drainage Board (MWSDB) now able to raise its own loans and with the power to operate largely independently of the State government.[16] Melbourne's MMBW had extensive powers to make bylaws and expand water supply catchment areas with the consent of affected councils, and could borrow up to £5 million.

[16] Hector, 'Sydney's water sewerage and drainage system', 18.

Perth's Metropolitan Water Works Board, established in 1896 to administer the works purchased from a private company, was abolished six years later, its responsibilities transferred to the Public Works Department and, from 1910, to a Department of Metropolitan Water Supply, Sewerage, and Drainage. A 1927 Royal Commission found Brisbane's Metropolitan Water Supply and Sewerage Board, established in 1911, to be ineffective and recommended that it be taken over by the City Council. In Adelaide, water infrastructure remained under the control of the State government Department of Engineering and Water Supply.

After Federation, State governments drew on financial resources obtained through their power to levy income tax to fund projects and provide subsidies for water authorities.[17] Public investment was increasingly directed to metropolitan areas, as State governments responded to demand from citizens to rent or purchase detached dwellings in new suburban subdivisions. The two wealthiest States, New South Wales and Victoria, could afford to electrify their suburban railway systems, and suddenly many more households were demanding water connection to homes established further from city centres. Municipal governments, with responsibility for building and maintaining local roads, accounted for an increasing share of public investment.[18] In the five years from 1919 to 1920, water and sewerage accounted for an average of 12.5 per cent of total public investment, around half that of railways. By 1938–39, the share of water and sewerage averaged 17 per cent over five years, close to that of railways.[19]

The establishment of metropolitan water authorities was a product of the practical difficulties of co-ordinating infrastructure investment of urban areas as a whole within a framework of fragmented local government. In Melbourne, hopes that the MMBW would develop into a 'parliament of suburbs' that would provide a means for citizens to influence policy or promote debate across municipal boundaries were soon dashed. 'Instead it became a rival, a juggernaut with a life of its own, liable to career off-track without warning.'[20] Responsibility for water management passed into state hands, in the form of water authority engineers and planners. In Adelaide, Perth, and, after 1928, Brisbane, citizens had a voice on water management issues through pressure on elected

[17] Commonwealth income tax was first levied during World War I. In the interwar period, both the Commonwealth and States levied income tax, with these powers transferred to the Commonwealth under Uniform Income Tax legislation in 1945.

[18] Sinclair, 'Capital formation', 38.

[19] Butlin, Barnard, and Pincus, *Government and Capitalism*, 34.

[20] Dingle and Rasmussen, *Vital Connections*, 105.

representatives, but in Melbourne and Sydney, powerful boards could only be influenced through municipal conferences, calls for inquiries, and criticism in the press. Private developers, builders, and households influenced water authorities indirectly by building, buying, and living in new suburban areas, creating demand for water and sewerage. The ability of authorities to meet that demand would depend on geographic and environmental constraints, the legacy of past infrastructure investment, political influences on the availability of resources, and the effectiveness of central planning.

The extension of sewerage lagged behind that of reticulated water because of its higher cost and the availability of cheap substitutes (septic tanks and pan toilets) that landlords and owner occupiers could install while waiting for sewerage connection. As metropolitan areas expanded, houses were built further from ocean outfalls, and growing volumes of waste water required augmented sewer capacity and expensive treatment facilities at discharge points.[21] A post–World War I house construction boom set authorities a challenge in keeping up with demand for sewerage, although the task was eased by suburban development being concentrated along public transport arteries. The servicing of established urban areas created expectations of similar standards from homemakers in new suburbs, as Dingle and Rasmussen note: 'A generation had now grown up in sewered houses and did not want to build in unsewered, let alone unwatered, areas.'[22]

THE DEMAND FOR WATER AND SEWERAGE

The continuing growth of the five metropolitan areas increased demand for water while supplies dwindled. They added almost a million inhabitants from 1901 to 1921, 41 per cent of them in Sydney and 29 per cent in Melbourne (see Table 4.2). Total population increased by almost 65 per cent – from 2.3 million to almost 3.8 million between the censuses of 1921 and 1947.[23] In response to this hefty growth, the cities made total net additions of 632,000 to housing stock. This represented an

[21] Walder, 'Water supply and sewerage in an expanding metropolis'.

[22] Dingle and Rasmussen, *Vital Connections*, 129.

[23] Merrett, 'Australian capital cities in the twentieth century', 190–91. Sydney and Melbourne each added more than a million inhabitants; Adelaide, Brisbane, and Perth close to half a million in total. Natural increase contributed 43 per cent of the growth of the five cities between 1911 and 1947 and intra-state migration 42 per cent, as people moved to the capital cities from provincial towns and rural areas.

TABLE 4.2. *City populations (thousands), 1901–47 (1); percentage change per annum over census period (2); ratio of percentage increase in housing stock to percentage increase in population over census period (3)*

	Sydney			Melbourne			Adelaide			Brisbane			Perth		
	(1)	(2)	(3)	(1)	(2)	(3)	(1)	(2)	(3)	(1)	(2)	(3)	(1)	(2)	(3)
1901	496			478			141.4			119.4			56.8		130.5
1911	648	3.1	125.9	593	2.4	111	169.2	2	191	139.5	1.7	N/A	107.1	8.9	92.9
1921	899	4.3	115	766.5	3	125.9	255.5	3.5	125.6	210	5	118	154.9	4.5	153.2
1933	1235.3	3.1	152.4	991.9	2.4	150	312.6	1.9	172.2	299.8	4	136.4	207.4	2.8	126.9
1947	1487	1.4	155.9	1226.4	1.7	146.2	382.5	1.6	135.4	402	2.4	127.9	272.5	2.2	130.5

Source: For 1901, McCarty, 'Australian capital cities'; for 1911–47, *Census of Commonwealth of Australia*.

increase of 191 per cent, compared to a population increase of 127 per cent.

Melbourne, Adelaide, and Perth were already sprawling cities, with decentralised economic activity due to the separation of their centres from port towns and a high proportion of the population living in detached cottages and villas.[24] Adelaide and Melbourne built extensive private tramway networks that operated efficiently on wide, generally flat streets. Adelaide had become overbuilt in the early 1880s, with excess capacity in housing and public transport, as was Melbourne later in the decade. Sydney and Brisbane were more compact, with most housing within walking distance of central workplaces. In Sydney, harbour ferries serviced the eastern suburbs and north shore. Suburban development was constrained by the distance of the rail termini (at Redfern) from city workplaces, the lack of an effective tramway network, and the ongoing unwillingness of a rural-dominated New South Wales parliament to invest in capital city infrastructure.[25] Brisbane's main streets were half the width of Adelaide's and its central grid was hemmed in by the Brisbane River.[26] After the depression of the 1890s, economic recovery was based on exports of wheat and other primary products and the growth of the manufacturing sector, due to import-replacement encouraged by population growth and protective tariffs. Perth was the fastest growing city in 1901–11; the railway to Fremantle provided a corridor for suburban expansion.

Before World War I, the prosperity of the five cities was contingent on the servicing, processing, and transport of resource-based output from their rural hinterlands. Government policies such as railway building, investment in infrastructure, and assisted immigration had stimulated urbanisation indirectly, but after the War governments became more active in their attempts to increase the level of urban employment through tariffs and investment in large-scale power generation and transmission. By the start of World War II, the cities had become their own hinterlands. Demand and supply shifts re-shaped the economy, with production costs reduced by cheap energy and labour-saving technology and by new industries emerging to meet the demands of suburban consumers, such as motor vehicles and consumer durables.[27]

[24] Frost, *The New Urban Frontier*.

[25] Wotherspoon, 'The "Sydney interest" and the rail'.

[26] Freestone, *Urban Nation*, 111.

[27] Merrett and Ville, 'Tariffs, subsidies, and profits', 58.

The cities' suburban footprint expanded, following public transport routes, raising the costs of providing water infrastructure. The population of Sydney's old core and inner suburbs, which in 1881 had housed 85 per cent of the metropolitan total in an area of 36 square miles (93 square kilometres), had almost peaked by 1911 and would grow by only 16 per cent to 1921. Between 1911 and 1921, Sydney's suburbs increased their population by 313 per cent and their share of the metropolitan population from 28 to 58 per cent, creating new demand for housing and water infrastructure.[28] Population growth resumed after the War, and by 1920 there were reports of a Sydney 'house famine', with city and inner suburbs balconies enclosed to house lodgers.[29] The eastern suburbs were already built up, with sewerage, but suburban growth in other parts of the metropolitan area took homemakers to sites where there were no sewers and, in some cases, no piped water.[30] Melbourne's public transport network was so extravagantly built during the 1880s that even after World War II the city still had large areas of potential suburban land within reach of fast commuter railways. In Melbourne, Adelaide, and Perth, stretches of farmland and bush awaited residential development. Electrification of Melbourne's suburban rail network began after World War I and was completed in 1924, allowing people to live further from the city without an increase in commuting time. After the War, a public authority, the Melbourne Metropolitan Tramways Board, took over the private and council tram systems and electrified, connected, and extended old cable and horse tram routes.[31] Sydney's network of suburban tramways was electrified, the railway system extended, and a new Central Station opened in 1906. The underground rail loop (1926), Harbour Bridge (1932), and tramway extensions provided more distant suburbs with better access to the inner city.[32] Most of Sydney's population growth in the 1920s was located within around 5–20 km of the city centre, especially in western and south-western suburbs such as Canterbury and Bankstown.[33]

As urban historian Robert Fishman observes of pre–World War II Los Angeles, electrified mass transit systems that extended over large metropolitan areas brought freedom to escape congested inner-city areas but

[28] Coward, *Out of Sight*, 222; Commonwealth of Australia, *Census*, 1911, 1921.

[29] Spearritt, *Sydney's Century*, 31; Sinclair, 'Capital formation', 32–33.

[30] Spearritt, *Sydney's Century*, 38–42.

[31] Ward, 'Development of Melbourne', 222–35.

[32] Kelly, 'Pleasure and profit'; Gibbons, 'Fall of the giant'.

[33] Spearritt, *Sydney's Century*, 37–40.

also imposed discipline on commuters and the built environment.[34] As in Australian cities, people and homes could spread out but had to remain within walking distance of transit stops. This encouraged the formation of coherent local neighbourhood centres, with their own identity and commercial and entertainment facilities near stations and stops that were linked to a downtown hub or Central Business District. Beyond walking distance, land remained undeveloped. As real estate speculation and public transport improvements stretched the metropolitan physical fabric, industrialists took up large sites that were integrated into the suburban landscape. In these largely unplanned, human-scale communities, the costs of providing water and other infrastructure were lower than if houses and business activity were dispersed across large areas.

Ornamental gardens were important markers of suburban respectability in Australian cities. From the nineteenth century, lawns served practical purposes as a means of controlling sand and dust, creating outdoor spaces for cooling, and enhancing the visual appeal of houses – but were heavy users of water.[35] In the five cities, water supplies that were sufficient for summer garden watering were expected, as a writer in the Melbourne *Argus* contented in 1910:

> To the citizen – the householder, with his suburban plot of thirsty garden to be watered – the coming of summer conjures up . . . days gone by when but the merest trickle scarcely sufficient to disturb the over-wearied and resting flies, was all his bath shower could boast in the morning; while his evening homecoming was saddened by the sight of a drooping and parched border of zinnias and asters in the front garden, and a row of wilted cabbage seedlings at the back. He remembers too that it was not until well onto midnight that the garden house could be induced to afford his moribund flowers and vegetables temporary succour.[36]

Suburban gardens increased the demand for water and created new concerns over the adequacy of supplies. At the turn of the twentieth century, per capita water consumption in Sydney was 200 litres per day. By the mid-1940s, the figure had risen to 318 litres per day.[37]

[34] Fishman, 'Re-imagining Los Angeles', 252. [35] Gaynor, 'Lawnscaping Perth'.

[36] Cited by Dingle and Rasmussen, *Vital Connections*, 119.

[37] Davison, 'Down the gurgler', 51; Dingle and Rasmussen, *Vital Connections*, 400; O'Hanlon and Spearritt, 'From water engineers to financial engineering', 101. At the end of World War I, per capita consumption in Melbourne was around 250 litres per day; by the end of World War II, consumption had increased to 320 litres per day. Average daily consumption increased by 70 per cent between 1925 and 1935, while population increased by only 8 per cent.

BUILDING DAMS

In 1901–02, Sydney experienced a drought that 'brought the metropolis perilously close to complete water famine – closer than it has ever been since', as historian W. V. Aird wrote in 1961.[38] The Upper Nepean Scheme, comprising weirs on the Cataract and Nepean rivers to divert water to a storage reservoir at Prospect, was estimated to be capable of supplying a population of 540,000 when it was completed in 1888. When the level of Prospect Reservoir dropped to below the limit of gravitational flow in 1902, the metropolitan population was 523,000. Garden hoses and, later, the use of watering cans were banned. A Royal Commission recommended the construction of a storage dam on the Cataract River, which was completed in 1907. Further dams on the Cordeaux and Avon rivers were under construction in 1923, when drought caused supplies to fail in some areas of Sydney, necessitating water restrictions. 'It is quite probable that the resources of the present catchment will be exhausted before the additional reservoirs are constructed', observed a writer in the *Sydney Morning Herald*, 'and the position which has existed for the past twenty years might be repeated later on'.[39] MWSBD engineers calculated that when full, the Prospect, Cataract, Cordeaux, and Avon reservoirs could provide a daily supply of 0.3 million cubic metres, but this provided 'no margin of safety' – daily consumption averaged 0.27 million cubic metres and more than 0.34 million cubic metres on summer days.[40] The Board responded to this supply uncertainty by investing heavily in headworks, working to a rule of creating storage capacity that was sufficient to meet the city's needs during its longest recorded drought plus one year.[41] As the Cordeaux and Avon reservoirs neared completion in 1925, work began on a storage dam on the Nepean, with a dam at Warragamba Gorge to follow, planned for completion by 1938.[42]

In all five cities, the case for constructing large storage reservoirs was based on the limited capacity of works that drew water from running streams. Periods of maximum demand for water were likely to coincide with warm, dry weather conditions that reduced supply, increasing the likelihood of shortages, restrictions, and possible failure. Perth's water

[38] Aird, *The Water Supply, Sewerage and Drainage of Sydney*, 25.
[39] 'Water supply', *Sydney Morning Herald*, 16 December 1924. [40] 'Water supply'.
[41] Butlin, Barnard, and Pincus, *Government and Capitalism*, 246.
[42] Hector, 'Sydney's water sewerage and drainage system', 14.

supply was especially precarious in the early 1920s.[43] Despite the construction of service reservoirs throughout the suburbs to feed dam water by gravity to reticulation mains, population growth outstripped the capacity of the system. The 30-year-old Victoria Dam, built to supply 25,000 people, was now expected to serve six times that number. Some of the system's iron and steel pipes were corroded and too narrow for adequate supply. The use of garden hoses was banned except for a few hours between 5 am and 8 am. In summer, brown water trickled from taps, dubbed 'liquid sausage meat' by one correspondent, and supplies failed to reach houses in new suburbs.[44] 'On Sunday last', wrote a minister whose rectory was on the lower reaches of the Nedlands hill, 'there was no water from 6 am until about 8 pm. You can just imagine how we fared on a blazing day like that.'[45]

Furthermore, water in existing storages was more likely to be contaminated when levels were low during dry periods. New storages and distribution works that would raise levels would reduce the need for expenditure on water treatment. Prior to modern treatment, the only way to obtain pure and safe water was to separate sources from sewers and drains and reserve catchments free of habitation. Logging continued to be permitted on Brisbane's Crown Land water catchments, which were so large that acquisition and depopulation was impractical. In response to complaints of contaminated water during droughts in the 1870s and 1880s, the Board of Waterworks made early experiments with chemical analysis and aerating and screening at service reservoirs.[46] Adelaide residents were subject to water restrictions throughout the 1920s; regular complaints were made about the quality of water and the need for filtration.[47]

Melbourne's water supplies were the most secure, with large forest catchments that had been protected from settlement and other economic activity since the early 1870s.[48] The Maroondah aqueduct, opened in 1891, supplied water to service reservoirs in winter, when stream flows were at their peak, allowing Yan Yean to fill for summer use. A diversion weir and aqueduct on the O'Shannassy River was completed in 1914. These works did not exempt Melburnians from periodic water shortages

43 Gregory, 'A spirit of Bolshevism', 85.
44 'To the editor', *Daily News* (Perth), 16 March 1923.
45 *Daily News* (Perth), 7 February 1923.
46 Cossins, 'One hundred years of Brisbane's water supply', 23–25.
47 Hammerton, *Water South Australia*, 214–15.
48 Frost, *An Environmental History of Australian Rainforests until 1939*, 158–59.

and restrictions during droughts. In the early 1920s, work began on three major storages (the Maroondah, O'Shannassy, and Silvan reservoirs) that were projected to provide sufficient water for an estimated 2.4 million people at current rates of consumption, a figure the city was not projected to reach until after 1961.[49]

Brisbane faced a dual environmental challenge of securing water supplies and mitigating floods.[50] Water authorities were hampered by inadequate headwork storage, and despite investment in pumps from the Mount Crosby weir on the upper Brisbane River and service reservoirs throughout the suburbs, the city remained vulnerable to water shortages during droughts. In 1912, the Metropolitan Water Supply and Sewerage Board began construction of Lake Manchester Dam on Cabbage Tree Creek, a tributary of the Brisbane River; the work was completed in 1916, but the catchment area was too small to harvest much water in dry periods, and the reservoir would take five years to fill. After a further drought in 1923–24, the Board began searching for dam sites on the Stanley River, a major tributary with a higher yielding catchment than that of the Brisbane River.

Schemes to control the run-off of the Brisbane and Stanley rivers while also increasing water supply had been proposed since the floods of 1893, when cyclonic rains over several weeks raised the water level at the Brisbane General Post Office (GPO) to 8.35 m.[51] This followed recent flooding in 1887, 1889, and 1890. The Brisbane River also flooded in 1908, 1916, and 1922.[52] After further flooding in 1927 a Royal Commission was appointed, which recommended damming the Stanley as a matter of urgency to limit future flood damage.[53]

When the Brisbane City Council was established in 1925, amalgamating all municipal councils with a ten-mile (16 km) radius, the Queensland economy was in a downturn due to drought and stagnation in the pastoral and mining sectors. When the Great Depression began in 1929, Queensland was already in recession, and the economic decline was not as

49 Dingle and Rasmussen, *Vital Connections*, 132–41.

50 Cossins, 'One hundred years of Brisbane's water supply', 28–34.

51 Cook, *A River with a City Problem*, 22–43. A subsequent government inquiry revealed that a flood in 1841 reached a slightly higher level, 8.43 m.

52 Australian Government. Bureau of Meteorology, Known floods in the Brisbane & Bremer River basin. www.bom.gov.au/qld/flood/fld_history/brisbane_history.shtml, accessed 14 April 2021.

53 Cook, *A River with a City Problem*, 63–65.

precipitous as in other States.[54] After taking responsibility for water supply in 1928, the City Council lacked the resources to begin dam construction, but after a series of floods in 1927, 1928, and 1929 and the inundation of Brisbane with 3.3 m of water in 1931, the State government addressed the issue of flood mitigation. All government borrowing in Australia was controlled by the Loan Council, a voluntary Commonwealth–State organisation established in 1923 to coordinate state loans and debt reduction. At the 1932 Premiers' Conference, Queensland's Labor Premier, William Forgan Smith, proposed a vigorous public works policy to restore employment, nine months in advance of Franklin D. Roosevelt's inaugural address. The Premiers' Conference agreed to issue a national recovery loan of £15 million for Australian States to use for unemployment relief work.[55]

To manage its public works program, the Queensland government established a Bureau of Industry in 1933. The following year, a sub-committee of the Bureau recommended the construction of Australia's first dual purpose dam, to be used to about half capacity for water supply storage, with the remaining capacity available to regulate river flow. The dam would be at its lowest minimum supply in summer, when the flood risk was high, with the level allowed to rise at the end of summer and gradually reduce over winter. Obtaining probable flood contours and valuations of houses in flood-prone areas, the sub-committee concluded that while it was impractical to build a dam that could provide protection from a flood at the 1893 level (the estimated cost was £3.5 million), a smaller dam could be justified by the value of reducing the height, and therefore damage, of future floods.[56] Construction was delayed by World War II, but Somerset Dam began supplying Brisbane from 1946 and was completed in 1959 (Fig. 4.2).[57] This conservative approach to flood mitigation would prove inadequate in the face of the devastating floods of 1974.

[54] Evans, *A History of Queensland*, 179–80.

[55] Carroll, 'William Forgan Smith', 408–9; Schedvin, *Australia and the Great Depression*, 329–30. Roosevelt first used the term 'New Deal' at the 1932 Democratic National Convention, at roughly the same time as the Premiers' Conference. Forgan Smith also anticipated the idea of John Maynard Keynes that borrowing to pay for public works was investment, not expenditure, which would pay for itself through taxes and the effects on spending of the newly employed. See Wapshott, *Keynes/Hayek*, 154–61.

[56] Bureau of Industry, *Report on Recommendations by the Special Committee Appointed to Investigate and Report upon Brisbane Water Supply and Flood Prevention* (Brisbane: Government Printer, 1934).

[57] Cook, *A River with a City Problem*, 63–70.

FIGURE 4.2. Somerset Dam during construction, 1949 (State Library of Queensland).

Perth's water crisis prompted robust press coverage and complaints from the public. In 1920, local government representatives met with the Minister and Department engineers to discuss a major dam on the Upper Canning River. No funding was committed, as the government was heavily in debt, having borrowed £2.5 million for the Goldfields Water Supply Scheme.[58] At a rowdy public meeting in March 1923, Premier Sir James Mitchell announced the Hills Water Supply Scheme – diversion dams on the Upper Canning and Wungong Brook, and two storage dams, one on Churchman Brook to be completed as soon as possible, with the other on the Canning to follow. Under legislation in 1924, water rates were increased and a loan raised to cover the cost of the works. By 1930, programs of sustenance relief – part-time, rationed work for unemployed men funded by State income tax and later loans and Commonwealth grants – made cheap labour available for the Canning Dam.[59] Workers began to clear the dam site in 1930, and construction began three years later. Churchman Brook was completed in 1929; after water began

[58] Gregory, 'A spirit of Bolshevism', 85–87.

[59] Butlin, Barnard, and Pincus, *Government and Capitalism*, 182; Hunt, *Water*, 76.

flowing from the Canning in 1940, Perth would face no significant threat of water shortages for the next three decades.

The South Australian government approved the construction of the Mount Bold Reservoir, to divert water to the Happy Valley Reservoir closer to the city, once loan funds were available in 1931. Completed in 1937, the reservoir would almost double Adelaide's storage capacity, sufficient to meet water needs for an estimated thirty-two years.[60]

In 1934, Sydney experienced the beginnings of a drought that would extend for eight years. After good rains in 1942, dry weather continued as the 'World War II drought' (1937–45) affected eastern Australia.[61] The MWSDB imposed severe water restrictions in May 1940 that remained in force for the next two-and-a-half years.[62] These restrictions, and delays in completion of the Nepean and Woronora dams, resulted in the Board being 'exposed to public criticism that it did not relish (and which, to a large extent, it had not earned)'.[63] The 50-foot (15.2 m) high Warragamba weir, completed in 1940, was an emergency measure to supplement the existing Upper Nepean storages, with a pumping station and pipeline to Prospect Reservoir. In 1941, Board engineers began planning a dam on the Warragamba Gorge, about one mile (1.6 km) upstream from the weir. With a storage capacity more than three-and-a-half times that of all of Sydney's existing storages (and an area one-and-a-half times that of Sydney Harbour), Warragamba Dam was planned to be able to supply 4 million people – almost twice Sydney's population at the time the dam was completed in 1960 (Fig 4.3).[64]

SEWERING THE SUBURBS

In all five cities, sewage was expelled into bodies of water. In Sydney and Brisbane, where ocean and river outlets were available and publicly acceptable as sinks, sewage was largely untreated until water pollution became a political issue in the 1960s and 1970s, as we discuss in

60 'Mt. Bold Reservoir bill in House', *Advertiser*, 4 November 1931; 'Mt. Bold water for Adelaide', *Advertiser*, 16 November 1935.

61 Australian Government, Bureau of Meteorology, 'Previous droughts', www.bom.gov.au/climate/drought/knowledge-centre/previous-droughts.shtml, accessed on 14 April 2021.

62 O'Hanlon and Spearritt, 'From water engineers to financial engineering', 101.

63 Butlin, Barnard, and Pincus, *Government and Capitalism*, 246.

64 Sydney Water Board, *The Warragamba Project* (Sydney: Metropolitan Water, Sewerage and Drainage Board, 1957), no pagination; Merrett, 'Australian capital cities in the twentieth century', 172. World War II delayed the start of construction until 1948.

FIGURE 4.3. Warragamba Dam in 1960 (National Archives of Australia).

Chapter 7.[65] The main sewers of the City of Sydney and the contiguous eastern and western suburbs drained to an ocean outlet at Bondi. The southern and outer western suburbs sewerage system drained to a sewage farm at Botany. The Botany farm was low-lying and sandy, and as the city's population grew, complaints increased from residents about the smell of sludge ploughed into fallow beds and effluent drained into the Cooks River. In 1905, the MWSDB recommended closing the farm, and from 1916 sewage from the southern and western suburbs was pumped to an ocean outlet at Malabar.[66]

Melbourne's sewage farm at Werribee was more successful. Wastewater was directed to grazing paddocks before the treated water was discharged into Port Phillip Bay. The smell of sewage at Werribee was initially problematic, but this was addressed with extra water flushed through the system as more properties were connected. Farm staff

[65] Butlin, Barnard, and Pincus, *Government and Capitalism*, 246.

[66] Hector, 'Sydney's water sewerage and drainage system', 13; Coward, *Out of Sight*, 131–32. From 1926, northern suburbs sewers were diverted to an outlet at North Head.

developed a system of grass filtration that enabled greater volumes of sewage to be effectively disposed of on smaller areas of land.[67]

Work on Brisbane's sewerage system began in 1916 but was delayed by World War I. When completed in 1924, the main gravitational sewer ran 12 km from Toowong, an inner suburb west of the city centre, in a north-east direction to the riverside flats at Eagle Farm. From there, sewage was pumped to a sludge treatment works at Luggage Point, at the mouth of the Brisbane River.[68] Heavy rain could cause the sewers to be inundated, with raw sewage discharged into waterways before reaching the treatment works. By 1933, the Brisbane City Council's Water and Sewerage Department reported an annual loss of almost £142,000, with loan indebtedness of over £8.4 million. Interest payments accounted for 71 per cent of expenditure, while the slow rate of sewerage extension limited the growth of rate revenue. After fourteen years of construction, the system served only 32 per cent of the City's built-up area – at that rate, it would have taken thirty years to complete the system, assuming no further population growth.[69] Many of Brisbane's unemployed found relief work in the construction of sewers and roads – over 7000 houses were connected to from 1934 to 1937.[70] The Council operated a nightsoil collection service for areas that were without sewerage, operating several sanitation depots, including one at Luggage Point, where nightsoil was loaded from trucks onto barges and then emptied at sea three times a week.

In Adelaide and Perth, sewage underwent treatment before discharge. Although Adelaide's stormwater drains ran directly to the Torrens and were not connected to sewerage pipes, heavy rain could cause the discharge of raw sewage into the main drain leading to the North Arm of the Port River, an area of Port Adelaide that was prone to flooding. The sewage farm system relied on the constant presence of vegetation in order to transpire and purify the sewage, and as Adelaide grew in size, the farm had to be expanded, with newly acquired areas being levelled, drained, and fenced. The farm was the subject of complaints from neighbours, which intensified with the development of manufacturing and housing around the farm. In 1921, a Select Committee was appointed to inquire into the farm with a view to moving it further to the north.[71] Four years

[67] Dingle and Rasmussen, *Vital Connections*, 86, 162. [68] Cole, *Shaping a City*, 28.
[69] 'Take 30 years to sewer Brisbane', *Courier-Mail* (Brisbane), 25 November 1933.
[70] Greenwood, *Brisbane 1859–1959*, 503–06.
[71] 'The sewage farm', *Advertiser*, 2 July 1921.

later, the Commissioner of Public Works reported on surveys being undertaken on treatment methods based on the activated sludge system.[72] Sewage treatment works at Glenelg and Port Adelaide, based on an air-diffusion purifying process, opened in 1933 and 1935.[73] World War II delayed plans for replacement of the overloaded Islington farm. With no prospect of further expansion and minimal maintenance in the face of a growing population (and thus sewage load), the farm could not operate efficiently, and disposal of raw sewage into the outfall channel was frequent in the 1950s.

The treatment filter beds and river outfall at Perth's Burswood Island, constructed in 1906, were the subject of complaints about smells as properties were connected and sewage load increased. Reports of algal blooms and fish kills were further evidence of water pollution. From 1912, Fremantle City Council operated its own sewerage system, with three septic tanks and effluent pumped to the ocean. The same technique would be used at the Water Supply, Sewerage, and Drainage Department's treatment works at West Subiaco, completed in 1927. Unemployment relief work helped the Department to make further connections to areas such as Claremont and Cottesloe. Capital expenditure on water and sewerage was the largest item in Western Australia's capital expenditure account in 1935–36.[74] The Burswood Island filter beds were by that time no longer operational.

To help meet the costs of extending sewerage to new suburbs, the boards in Sydney and Melbourne relied on municipal guarantees to supplement the rate revenue needed to make loan repayments. From 1892, the Sydney Board required the councils to be served by new infrastructure to provide a guarantee bond or deposit, ranging from 4 to 8 per cent of the capital invested. After 1919, councils could levy special rates when bonds were required. The criteria were progressively relaxed, until in 1937 the State government guaranteed the first 2.5 per cent return, leaving councils responsible for the remainder.[75] The MMBW worked to a '6 per cent rule', only extending sewerage and water reticulation where councils guaranteed to pay any difference between a minimum 3 per cent return on rates to connected properties and 6 per cent.[76]

[72] 'Adelaide's sewage system', *Advertiser*, 28 July 1925.

[73] Hammerton, *Water South Australia*, 214–18. [74] Hunt, *Water*, 76.

[75] Butlin, Barnard, and Pincus, *Government and Capitalism*, 247.

[76] Dingle and Rasmussen, *Vital Connections*, 125.

In all five cities, landlords and owner occupiers were responsible for the cost of laying pipes and installing fittings to homes. Under Sydney's *Sewerage Act* of 1880, owners or occupiers were liable for water or sewerage rates if their land was within 150 feet of a water or sewerage main. In 1924, Parliament increased the rateable area to within 250 yards of a water main and 250 feet of a sewer main; an identical provision was included in Victoria legislation that year.[77] Once the MMBW laid sewers, the area was gazetted as a Sewer Reticulation Area, with all properties required to be connected within one month. An owner could submit a plan for connection and hire a plumber to do the work, or the Board could complete the work and bill the owner. These charges did not vary by property value and so were regressive in nature. The average connection cost of £20 was equivalent to six to ten weeks' wages for a labourer; three-quarters of households opted to spread payments to the Board over ten years.[78] 'I had to raise a small loan in order to have the place sewered', a Perth resident of the 1920s recalled. 'It cost me something over £100.'[79]

The work of the MMBW was complicated by tension between areas that were already served with infrastructure and those on the suburban frontier. Addressing water infrastructure problems required collaboration between a growing number of municipalities, as well as large amounts of capital. As councils across the metropolitan area resisted general rate increases, which were unpopular with ratepayers, Board members were more likely to postpone action than take on higher levels of debt. 'If the problem was diffused, intermittent, or merely aggravating – urban sprawl, poor roads, insufficient bridges, smelly drains or local flooding – there was little will to act.'[80] When the MMBW was established, outer suburban areas had resisted expenditure on sewerage, while inner suburban councils considered it essential. By the end of World War I, these positions had been reversed, as outer municipalities expected sewerage and campaigned to be included within the Board boundary (then set at 21 km from the Melbourne GPO), pressing for higher general rates to finance sewerage extensions. Outer suburban councils attempted to make nightsoil disposal a Board responsibility. Already sewered, inner suburbs now urged economy in Board operations.

[77] Coward, *Out of Sight*, 225; Dingle and Rasmussen, *Vital Connections*, 170.
[78] Dingle and Rasmussen, *Vital Connections*, 102.
[79] Mr. F. Miller, interviewed by Jenny Gregory, 1982.
[80] Dingle and Rasmussen, *Vital Connections*, 150.

During the 1920s, the MMBW supplied 80,000 new houses with water and connected 72,000 to sewerage. By 1931, 19,000 Melbourne houses had water but no sewerage.[81] The ability of the MWSDB to keep pace with Sydney's expanding metropolitan area was hampered by the development of suburbs that were distant from existing mains and an expanding area of operations that from 1916 included Wollongong. More than 58,000 houses became liable for sewerage rates between 1921 and 1931, but the percentage of unsewered houses increased from 23 to 26 per cent.[82] Keeping up with the demand for sewerage was much more difficult than keeping up with demand for water.

A CONTINUING CHALLENGE

The challenge for public authorities in the five cities during the twentieth century lay in maintaining momentum for networked urban infrastructure provision, keeping up with the demands of growing populations in cities that had safe drinking water and sewage disposal, and allowing the smaller cities to make up lost ground. Until the 1930s, fewer than half of Perth houses were connected to sewerage; Brisbane did not reach that mark until the 1960s.[83] By 1947, half of Brisbane houses relied on pan toilets – a figure comparable to that of American small towns (and much higher than 15–20 per cent in the four other Australian cities). The percentage of households in the five cities with running water (94–99 per cent) was comparable to that of US cities, but only in Adelaide and Melbourne were houses as likely to have flush toilets.[84]

New investment in water infrastructure was subsidised by ratepayers who were already served. Private suburban developers were thus able to shift part of their costs to ratepayers. The subdivision of suburban land, driven by demand for detached housing, was a lucrative business because councils did not require developers to provide fully made roads with gutters connecting to local drains. Although slowdowns in private construction during the Great Depression and World War II reduced the pressure of new housing that needed water services, shortages of capital funds also reduced investment in public infrastructure. Another boom

81 Dingle and Rasmussen, *Vital Connections*, 132.
82 Coward, *Out of Sight*, 226.
83 Butlin, Barnard, and Pincus, *Government and Capitalism*, 247.
84 Census of the Commonwealth of Australia, *Dwelling Summary Bulletins*, 1947; Gordon, *The Rise and Fall of American Growth*, 120.

after the War would create new demands. The proportion of Sydney houses without connection to sewerage was higher in 1956 than it was in 1901.[85] By 1947, sewerage was available to almost 45,000 Perth dwellings, two-thirds of the housing stock; by 1971, only 40 per cent of houses in the metropolitan area were sewered.[86]

85 Coward, *Out of Sight*, 225.

86 *Sewerage Connections and No of Dwellings in Perth Metropolitan Area at Census Years 1911–1976*, Australian Bureau of Statistics; S-J. Hunt, 'A history of water supply, sewerage and drainage in metropolitan Perth', unpublished manuscript (Perth: Metropolitan Water Supply, Sewerage and Drainage Board, 1979).

5

Transforming Homes

In 1895, James Brennan and his three brothers moved from Queensland to the Western Australian goldfields, operating a large drapery store in Kalgoorlie with branches in nearby towns. James moved to Perth when he married Elizabeth Fitzgerald in 1899 and, with his brother Patrick, operated his own store, the 'House of Brennan' on Hay Street. Twenty years later, James and Elizabeth bought a grand architect-designed, eleven-room, 6250 square foot (581 square metre) home on Kings Park Road, in the fashionable suburb of West Perth (Fig. 5.1). With additional land purchases in the early 1920s, its grounds were increased to over half an acre (2203 square metres). Constructed in 1907, the house featured a drawing room, dining room, smoking room, breakfast room (with adjacent servery with sink), kitchen (without a sink), scullery (with sink), and lavatory and cloakroom off the main corridor from an entrance hall on the ground floor. Upstairs, there were five bedrooms, but only one bathroom, with basin, bath, and water heater, and a lavatory on the other side of the adjacent stairwell. A large verandah wrapped around two sides of the house on the ground floor, as did a balcony on the second floor. Although the property would have been reticulated when constructed, water was also drawn from two external thousand-gallon (4546-litre) galvanised iron rainwater tanks. Four-inch pipes from the bathroom and lavatories drained to a gully and septic tank at a back corner of the lot. An outside toilet on the other corner of the back fence also drained to the septic tank. Drains for excess water led to a dry well near the septic tank.[1]

[1] Hobbs, Smith, and Forbes, Architects, Weld Chambers, St Georges Terr, Perth, Proposed residence Kings Park Road Perth, c. 1907, State Library of Western Australia,

FIGURE 5.1. Brennan home, West Perth, 1921 (*Sunday Times* (Perth), 18 December 1921). Note reticulation pipes and sprinkler on tennis court.

Such rudimentary plumbing amid spacious elegance is a stark reminder that only recently have water and sanitation become abundantly and conveniently available inside suburban homes. Homing in on this transition in the first half of the twentieth century shows how gender, class, and race crucially shaped interactions with water and its expanding infrastructure in Australian cities.

The Brennans' move to Perth at the turn of the century mirrored a nationwide trend, as Australians gravitated towards capital cities along the coast. Perth's housing market responded more slowly to early twentieth century population growth than the other four cities, as shown in the previous chapter. Between 1901 and 1947, Perth's share of the Western Australian population increased from 20 to 55 per cent.[2] It was the only Australian capital city where the housing stock grew more slowly than population during an intercensal period (between 1911 and 1921) but

ACC. 9592AD/64.2, https://purl.slwa.wa.gov.au/slwa_b4781084_1, accessed 8 February 2021; Western Australian Post Office Directories, 1907–39, www.slwa.wa.gov.au/explore-discover/wa-heritage/post-office-directories, accessed 16 March 2021.

[2] Merrett, 'Australian capital cities in the twentieth century', 175.

made up ground in the interwar period. Between 1921 and 1947, Perth's housing stock increased by 112 per cent, while population increased by 76 per cent. The number of persons per dwelling declined from 4.9 to 4.1 during this period. The expansion of suburbs transformed the environment of the Swan Coastal Plain, as bush was cleared to make way for new housing.

As in the USA, the gradual replacement of labour-intensive 'private-lot waste removal' methods by networked infrastructure made possible a change in Australian house layouts. Informal, mixed-use 'worker cottages' and 'villas' gave way to the 'bungalow', a small, simple dwelling with more specialised interior spaces and a bathroom adjacent to bedrooms, which allowed rising living standards to filter down to working-class families.[3] Housing styles changed with the emergence of the Californian bungalow, which with local variations was the dominant housing form in new interwar suburbs in Perth, as well as Sydney, Melbourne, and Adelaide. In Brisbane, the high-set or elevated 'Queenslander' style that emerged in the 1920s blended traditional cottage forms and materials with newer influences.[4]

The water use of suburban householders also underwent a dramatic change: in 1941, Perth residents consumed an average of 306 litres daily, nearly twice as much as before World War I.[5] Some of this increase in water use can be attributed to the increased domestic availability and accessibility of water, which resulted from the growing reticulation of the suburbs outlined in Chapter 4. The opening of the Kalgoorlie Pipeline in 1903 had created a disparity in comfort between those on the eastern goldfields and those in the suburbs of Perth, who remained largely reliant on supplies from artesian bores.[6] Even in the early 1920s, the local *Daily News* would opine, 'There is no sound reason why the people of the metropolitan area should be treated differently from the rest of the population.'[7] Reticulation was frequently used as an enticement for prospective residents of the new suburbs, and the Metropolitan Water Supply, Sewerage, and Drainage Board (MWSSDB) increased its water storages to ensure these supplies would be available.[8] By 1913, all areas serviceable by gravity had been reticulated. Yet, nearly a quarter of

3 Groth, 'Workers'-cottage and minimal-bungalow districts'; Gordon, *The Rise and Fall of American Growth*, 108–10.

4 Bell, *A History of the Queensland House*, 42–46.

5 Hunt, 'Address'.

6 Hartley, *River of Steel.*

7 'Water for the metropolitan area', *Daily News*, 17 May 1923.

8 Hunt, *Water.*

Perth's homes remained without piped water, dependent upon wells and neighbours' taps for their supplies. Many more continued to rely on galvanised iron tanks to supplement the expensive and unreliable reticulated supplies.[9] Even where homes were connected to mains water, there were disparities in the extent and type of indoor water and sanitary provisioning.

In their early stages of development, frontier towns conformed to a pre-industrial form, physically structured to minimise distances people had to walk to go about their daily business. The wealthy lived and worked in houses along principal thoroughfares, with the labour force living in alleys and backstreets. 'First-generation cities were a confusing hodgepodge', observes historian Carl Abbott of settler towns in the American West. As towns grew from an original locus such as a harbour or waterfront, 'nobody knew for sure where the "good side" of town was going to be'.[10] From the 1830s, the middle classes followed the trend established by London's merchant elite, moving to private villas in open countryside at the edge of the city to escape the increasing crowding, poverty, and pollution of city cores. The ability to afford transport sorted city populations along class lines, as whole neighbourhoods and suburbs assumed a more homogeneous socio-economic composition.[11] By the end of the nineteenth century, Melbourne, Sydney, and Adelaide had progressed through this phase; Perth and Brisbane had begun to do so.

As shown in Chapter 3, the spatial sorting of populations along class lines acted as a queuing device for water infrastructure. Affluent districts were generally the first to be connected, and poor ones suffered longer the effects of inadequate infrastructure. As in other Australian cities, Perth's elite suburbs developed close to a water supply. Proximity to the government-owned Mt Eliza Reservoir allowed developers in the late 1890s to advertise West Perth as 'the most fashionable and healthiest part of the city', 'with water laid on'; wealthy Peppermint Grove was served by the private Osborne Water Works.[12] But supplies were often unreliable in summer, and piped water, where available, was expensive; in 1903, the charge to private consumers in Perth was double that of Melbourne and Sydney.[13] Although working-class Subiaco was also reticulated before

[9] Stannage, *The People of Perth*.

[10] Abbott, *How Cities Won the West*, 133.

[11] Davison, *City Dreamers*, 40, 90.

[12] Kelly, 'The development of housing', 170. See for example 'Auburn: The Toorak of Perth', January 1896 (poster in possession of Jenny Gregory).

[13] 'Threatened water famine', *West Australian* (Perth), 4 January 1901; Stannage, *People of Perth*, 277.

1900, the cost of water mattered to those householders, and many continued to use tank water. After 1906, brick outhouses located against a back fence, with access from a right-of-way or lane for emptying of dry-earth or pan toilets by a contractor, became mandatory in new subdivisions.[14] Households that could afford septic tanks were able to include lavatories inside the house.

In focusing on water supply and the provision of sewerage to Australia's Edwardian, interwar, and World War II suburban housing, this chapter traces the changing configuration of houses with the gradual move to specialised rooms for baths and toilets, along with laundries. As architect Robin Boyd later put it, having knocked on the 'posts of the rear verandah', taps, drains, and water closets were now 'allowed to step up' into the house.[15] This process of sanitary stepping up shaped not only the evolution of Australia's small suburban homes but also the habits and aspirations of their residents, both indoors and out. The Australian ideal of private home ownership was paradoxically collective, as residents sought to maintain suburban standards of respectability that became increasingly reliant on the growing networks of water supply and sanitation across the suburbs. 'Nearly every man's private castle was tied to its neighbours', Boyd observed, 'and to every other private castle in town by two or more wires attached to its roof and by a more or less complicated tangle of pipes underground'.[16] These sanitary connections blurred private and public spheres, with lasting environmental, technical, and political implications. The linkage of households to pumping stations and dams, which allowed changes in private routines, became the basis for political mobilisation around the quality and reliability of domestic water services.[17]

The conjunction of these events in Perth provides a case study to enable us to test Robert Gordon's proposition, based on his analysis of historical trends in the United States, that the extension of network infrastructure had a democratic influence on living standards.[18] For low-income households in the five cities, fixed costs of sewerage connection and lavatory installation were a barrier. Tenants were at the mercy of landlords to provide such services.[19] Widening the gaze to the proliferation of Perth's infrastructure similarly suggests that the benefits of such expanded

[14] Kelly, 'The development of housing', 28. [15] Boyd, *Australia's Home*, 49.
[16] Boyd, *Australia's Home*, 244. [17] Taylor and Trentmann, 'Liquid politics'.
[18] Gordon, *The Rise and Fall of American Growth*.
[19] Merrett, 'Economic growth and well-being'; Macintyre, 'Owners and tenants'.

networks accrued unevenly across suburban Australia in the early twentieth century, which qualifies to some extent Gordon's egalitarian thesis.

THE GOLDEN WEST

Until the gold rush boom of the 1890s, rich and poor lived in close proximity in central Perth. Decentralised at birth due to the separation of port and town centre, Perth's suburban development gathered pace after the construction of railways to Guildford (1881) and Fremantle (1883). Subdivided from 1895, land in the city's west ward (West Perth) offered desirable residential sites for households wishing to escape increasing commercial activity, some close to the 1017-acre (412 hectare) Perth Park (opened in 1895 and later renamed Kings Park). Elevated and well drained, West Perth was close to the Swan River, with cooling summer afternoon sea breezes (the delightfully named 'Fremantle Doctor', as it was known by 1870).[20] As in genteel suburbs worldwide, the area was located predominantly upwind of the smells of the city. Linked to the city centre by the railway and an electric tramline, West Perth was home to professionals, senior civil servants, and parliamentarians.[21] At the 1901 census, nearly a quarter of West Perth houses had seven to ten rooms, and the suburb employed the largest number of servants in Perth.[22] Despite this affluence, West Perth remained a blend of socio-economic groups. One-third of houses in the suburb was of three or four rooms, many on smaller and narrower lots on lower ground adjacent to the Perth–Fremantle railway line.

The city's east ward (East Perth) was less of a residential suburb – by 1904, around half the land was used for housing, compared to three-quarters in West Perth.[23] East Perth was a site of early settlement, with elite homes and gardens on one-acre lots close to the river. The southern part of East Perth remained an area of respectable housing, as did the higher ground to the north, but from the mid-1890s, settlement spread to cheaper, low-lying, and damp land to the south-east. A tent city sprang up to accommodate prospective (or disappointed) goldminers near two open drains that ran along a trough from the City to Claisebrook Creek. Stables, tanneries, brickworks, workshops, factories, polluted drains, and cheap housing in the vicinity confirmed the social character of the

[20] Christensen, 'Fremantle Doctor'. [21] Wises Post Office Directories, 1903–1914.
[22] Kelly, 'The development of housing', 254; Stannage, *The People of Perth*, 243.
[23] Stannage, *The People of Perth*, 243.

area. In 1900, only a quarter of East Perth housing was owner-occupied; almost half of the housing stock was built for landlords holding at least four properties.[24] At the height of the gold boom, rents for worker cottages in Perth were reported to be twice those of Melbourne, Sydney, and Adelaide and around a quarter of unskilled wages.[25]

Perth was ringed with suburbs by 1911 – two in three of the metropolitan population lived outside the city centre. The opening of railway stations at Subiaco (1883) and Leederville (1897) on the Fremantle line and Bayswater (1896) and Bassendean (1910) on the Guildford line signalled the emergence of new artisan and working-class suburbs. On the western line, an industrial area developed at Subiaco, and on the eastern line, jobs were available at the Railway Workshops at Midland and superphosphate plants, steel works, and other industries at Bassendean.[26] Suburban development in Perth in the interwar years was fed by natural increase, migration from rural areas, and immigration from Britain (the overseas immigration component of Perth's population growth from 1911 to 1947 was the highest of any of the five cities – more than double that of Sydney).[27] Access to transport, the location of industry, topography, and the consequent price of land meant that Perth suburbs increasingly reflected a pattern of residential segregation by class and race. The level of home ownership in interwar Perth was very high: increasing from 41 per cent in 1911 to 55 per cent in 1921 and to 72 per cent in middle-class suburbs such as Nedlands.[28]

Perth's suburban expansion increasingly encroached on the lifeways of local Aboriginal peoples, who found on the city's outskirts places to camp and pick up work while maintaining connections to Country. A group of about twenty Whadjuk Noongar lived near Galbamaanup, the swamp on the other side of the railway in Claremont. Without access to the city's growing sanitary services, they obtained water by digging wells or, as one remembered, 'in an old kerosene bucket from a tap at the back of the police station' or from 'Neil's paddock close by because he had a pump there'. Another group of Aboriginal people camped each summer near a riverside beach on Freshwater Bay. They occasionally asked white residents for water but mainly sourced 'brackish but drinkable' water by 'digging a hole in the beach sand a yard or two from the high tide mark

24 Thomas, 'East Perth', 152.
25 Kelly, 'The development of housing', 194.
26 Fox, 'Work organisation'.
27 Merrett, 'Australian capital cities in the twentieth century', 190.
28 Gregory, 'The manufacture of middle-class suburbia', 172.

anywhere along the beach', as Corrie Bodney recalled.[29] Each morning before school, Robert Bropho remembered, 'There was no baths in them days for us. There was just a cat wash ... a couple of handfuls of water thrown over your face from an old kerosene tin cut down the sides to make it like a dish.'[30]

To limit interactions with such Indigenous peoples, the local council shifted the Aboriginal camp at Galbamaanup in 1912 'because of the fly nuisance'.[31] As historian Anna Haebich argues, 'Building on their racist attitudes towards Aborigines, the settlers used the Aboriginal presence as a useful negative definer of what the new communities believed they stood for; for example "cleanliness" and "orderliness" versus "disease" and "dirtiness."'[32] A Health Inspector at Hurstville Council in Sydney had been surprised in 1926 to report that weatherboard cottages owned by Aboriginal families were 'surprisingly clean and well-kept' with 'tidy' surroundings, despite the complaints of white residents that conditions there were unsanitary.[33] Confined to the margins of white settlements on reserves and in camps, Aboriginal people could be better regulated and controlled under what historian Gyan Prakash has described in nineteenth-century India as 'a discriminatory sanitary order'. Protecting the health of white suburban Australians, as in colonial India, required the containment of the putative source of disease – the bodies, habits, and homes of Aboriginal people.[34]

ANXIOUS SUBURBS

After World War I, concerns were reignited among the British and Australian middle classes about racial degeneration and the decline of the Empire.[35] Roused by these worries, reformers from an emergent class of technocrats sought to overcome social ills through the application of principles of science and reason to domestic life.[36] Stimulating this

[29] Ern Middleton, 'I remember Swan River Playground', undated newspaper cutting, Freshwater Bay Museum; Corrie Bodney, oral history recorded by Denise Cook, 24 February 2012; 'Our camp at Freshwater Bay', http://www.freshwaterbaymuseum.com.au/wp-content/uploads/2013/06/CM4_Noongar_F_web.pdf, accessed 27 March 2021.

[30] Bropho, *Fringedweller*, 8.

[31] Bolton and Gregory, *Claremont*, 119.

[32] Haebich, *For Their Own Good*, 129.

[33] Cited in Goodall and Cadzow, 'Salt Pan Creek', 202.

[34] Prakash, *Another Reason*, 132–33.

[35] Gregory, 'Protecting middle-class suburbia', 77.

[36] Reiger, *The Disenchantment of the Home*, 2–3.

movement were widely held anxieties about the influence of the environmental conditions inside and outside the (middle class) home on the mental and physical health of white Australians. The body and its health had come to be seen as a closed system, distinct and separate from its environment. These ideas point to the paradox that historian Linda Nash observes in modern public health: 'It insisted on the need for certain environmental changes while denying that the environment played an active role in the production of disease.'[37] In line with the modernist state, advocates of these environmental reforms were mostly male members of an emerging professional middle class seeking to reshape Australian suburbs, homes, and families according to their vision of a modern society.[38]

In Perth, local advocates such as William Saw, Billie Bold, Harold Boas, and Bessie Rischbeith called for the improvement of parks and playgrounds to better the lives of Western Australians, particularly children.[39] Water was an important yet often overlooked tool to achieve such reform. By the end of the nineteenth century, a constant water supply had come to symbolise modern civilisation, and these supplies provided the means to cultivate attractive open spaces, through reticulation of gardens and the installation of fountains.[40] In each of the five cities, fountains and water features, ranging from the modest to the grand, used the active ornamentality of moving water to inscribe meaning in public spaces, memorialising elites and celebrating the triumph of infrastructure over nature as emblematic of 'civilisation'.[41]

The expansion of Perth's reticulated water supplies had coincided with changing expectations about the appearance of the city and its suburbs. During World War I, Perth's town planning movement emerged, and its influence on suburban development in Western Australia was almost immediate. The leaders of the local movement widely advocated the importance of orderly planning, open spaces, and aesthetic appeal in town planning.[42] In an address to the Royal Society of Western Australia in 1918, Saw, the president of the Town Planning Association of Western Australia, argued:

> We must do better than we have done in conserving our baby life by taking greater care of the mothers of the nation ... by getting [them] ... out ... to zones ... where

[37] Nash, *Inescapable Ecologies*, 113. [38] Reiger, 'All but the kitchen sink', 497.
[39] Gregory, 'Let our watchword'. [40] Taylor and Trentmann, 'Liquid politics', 204–05.
[41] Gaynor and Martin, 'Every fountain tells a story'.
[42] Gregory, 'Let our watchword', 181–82.

the children, in their garden villages, will grow up taller, stronger, deeper in the chest, freer from physical defects, happier, more likely to be stalwart effectives in the wealth-creating forces of the State, and less likely to be a burden on the community.[43]

Attention to these details not only reflected the ideals of the international town planning movement but was also evidence of prevailing anxieties about the influence of the environment on the mental and physical health of suburban Australians.

Creating the ideal environment for the healthy moral and physical development of white Australian children was the duty of the state's citizens. Citizenship was therefore closely related to the upkeep of the home's outward appearance. The maintenance of the front garden in particular was, according to historian Robert Freestone, 'vital in order to attain the coherence of the garden suburb street picture, which secured a demonstrable sense of community'.[44] Garden fashions called for a large expanse of lawn, with flowering shrubs and annuals.[45] The backyard, in contrast, was a utilitarian space for household tasks, as well as food production.[46] To allow the front garden to deteriorate would undermine the bonds of the community and reflect interwar suburban anxieties of racial and imperial decline.[47]

These middle-class concerns about the appearance of the front garden led to the heavy use of water outside Perth homes, particularly during the long summer months.[48] During the period of water restrictions in the early 1920s, some local businesses appealed to these anxieties and encouraged gardeners to invest in windmills and engine pumps to 'be independent' and to ensure 'a free water supply, when you want it, and where you want it'.[49] It was only after this period of water restrictions that locally produced gardening publications, such as the *Western Australian Gardening Guide*, became available and advocated methods of water conservation. In 1925, one Perth author suggested that summer gardens would likely be around half the size of the winter garden.[50]

[43] Saw, 'Some aspects of town planning', 41.

[44] Freestone, 'Planning, housing, gardening', 133.

[45] Viska, *A Guide*, 26–27.

[46] Seddon, *Landprints*, 153; Gaynor, *Harvest of the Suburbs*.

[47] Gregory, 'Protecting middle-class suburbia', 77.

[48] D. R. Williamson, 'Statistics of water use', 6–7.

[49] Bros, 'Advertisement', *West Australian*, 23 October 1920; Malloch Bros, 'Advertisement', *West Australian*, 22 January 1921.

[50] Gaynor, *Harvest of the Suburbs*, 22.

Gardens are material, sensory projects, as well as ideological ones. In the suburbs of Perth, lawns and other garden plants were allies in the ongoing war against the sand and heat that invaded settler homes. They cooled the air that the Fremantle Doctor brought through the home on summer afternoons, and they trapped dust and sand under a green layer that not only evoked a long tradition of elite English landscapes but also provided a soft surface for children to play on and a place to work or talk with friends.[51] In 'lawnscaping' their homes, residents sought both amenity and status. These aspirations were articulated in a 1922 debate over the annual budget for the Metropolitan Water Supply, Sewerage, and Drainage Department. When the Minister for Works proposed that home gardeners used too much water, the Member for working-class South Fremantle defended them as 'trying to ... build up a metropolitan area approaching those elsewhere in the Commonwealth', instead of being 'allowed to live on the level of the aborigines who were here half a century ago'.[52] For white residents, lawns meant not only comfort and convenience but also an acceptable place within racial and imperial urban hierarchies.

Although Perth's gardeners were especially outraged at the restrictions on their water use in the early 1920s, many householders found their reticulated water supplies had literally dried up. In March 1920, the *Daily News* declared, 'Last night practically the whole city was without water, and the sorry spectacle was witnessed, even in Hay Street West, of mothers taking their pitchers to those who possessed wells, to obtain water with which to wash their children before putting them to bed.'[53] Reminiscent of the city's water famine of the late 1890s, this account highlights the domestic relationship between women and water and the ways in which gender relations shaped water use in suburban Australia in the early twentieth century. At the time of these shortages, nearly half the average Perth household's water consumption took place inside the home.[54] As the *Daily News* account suggests, such episodes of water scarcity interrupted the domestic routines of Perth's housewives and domestic servants, adding to their heavy household burden.[55] Shortages exposed the dependence of many suburban households on reliable scheme water supplies to maintain the veneer of civility and exacerbated suburban anxieties on the domestic frontier.

[51] Gaynor, 'Lawnscaping Perth'. [52] *WAPD*, 12 December 1922, 2161.
[53] 'A waterless city', *Daily News*, 14 March 1920.
[54] Williamson, 'Statistics of water use', 6–7.
[55] Evans and Saunders, 'No place like home'.

CALIFORNIA DREAMING

Adopted from the 'bangla' of Bengal, India, the 'bungalow' style was imported to Britain and the United States as a holiday house in the late nineteenth century.[56] Transmitted by pattern books with plans and articles in magazines and newspapers, the bungalow style spread quickly from southern California, being adopted almost simultaneously in Vancouver, the North American Midwest, and Australia. In 1907, *Building* (a Sydney-based 'magazine for the architect, builder, property owner and merchants') published images of bungalows designed the previous year by Greene and Greene, a Pasadena architectural firm that worked for affluent clients but hoped to set a pattern for mass housing.[57] A focus on this housing type provides insight into the lifestyle changes that were brought by the provision of water supply and sewerage.

By the early twentieth century, as rapid urbanisation increased the demand for housing, supply bottlenecks drove up the price of lumber. Architects and builders responded to rising costs by embracing the small (1000–1200 square foot, or 93–111 square metres) bungalow, which economised on space while bringing middle-class expectations of floorplans that offered formal separation of public, private, and functional spaces.[58] This informal style was a synthesis of diverse sources, with wedge-shaped brick pylons for deep verandahs (Spanish), minimalist roof framing with wide overhangs (Japanese), use of stained wood in construction and ornamentation (Swiss and Scandinavian), large windows with leadlight ornamentation, and easy access to outside areas, which suited the Californian (and Australian) climate and lifestyle. Built without cellars or expensive foundations, timber versions in North America 'were available for a few hundred to a few thousand dollars, and accessible therefore to the broadest range of house buyers'.[59]

After the end of World War I, most new builds in the five cities, including Perth, were Californian bungalows (Fig. 5.2). In Perth and Sydney, most were of brick with orange or red tile roofs; in Adelaide, the variation was a stone façade and corrugated iron roofing. In Melbourne working-class suburbs, weatherboard bungalows of five main rooms were common.[60] In Brisbane, houses retained the weatherboard

[56] King, *The Bungalow*.
[57] Clare, 'The Californian bungalow in Australia', 19; Starr, *Inventing the Dream*, 101–02.
[58] Doucet and Weaver, 'Material culture and the North American house', 583.
[59] Holdsworth, 'Regional distinctiveness in an industrial age', 70–71.
[60] Broome, *Coburg*, 210.

FIGURE 5.2. Californian Bungalow, Preston (Melbourne), c. 1932 (Museum Victoria).

construction, corrugated iron roofing, verandahs, and elevation on posts or stumps (with metal or treated caps to protect against termites) that characterised earlier two- and four-roomed cottages. In the 1920s, a 'Queenslander' style emerged that blended traditional elements with joinery and ornamentation and roofs that were lower pitched and dominated by gables, reflecting bungalow influences.[61]

In the Perth suburb of Nedlands, a new bungalow built in 1927 for the managing director of the new Perth branch of a South Australian hardware firm 'was more or less built as a display house' for the firm. The home boasted the latest in modern household conveniences:

> a stainless-steel sink, which was new, and jazzy taps, and a Junker's gas hot water system. Then there was the bathroom. ... We had mottled red tiles on the walls right up to the ceiling, as well as tiles on the floor. ... There was also a toilet in the bathroom, which was unheard of in those days. So we had two toilets, one outside as well as the one in the bathroom. And all the latest ceramic toothbrush holders. ... There was a bath and a separate shower, which again was most unusual, most bathrooms that had a shower had it over the bath.[62]

[61] Bell, *A History of the Queensland House*, 44.
[62] R. Harvey (pseudonym), interviewed by Jenny Gregory, April 1986.

Although few suburban households enjoyed such comforts in the early twentieth century, such was the growing appeal of domestic modernity that many aspired to the sanitary conveniences they could afford.

Home magazines of the interwar period, pitched to the growing middle class, frequently featured exterior and interior photographs and occasionally plans of a range of homes, both architect-designed and builder-designed. Photographs and articles about the bathroom increasingly appeared and gestured to the gendered nature of domestic work and cleanliness in middle-class interwar homes. As a 1929 article in the Perth magazine *Western Homes* began:

> Years ago, when the bathroom was not much more than a closet with a tub in it, the layout did not matter so much. Nowadays the bathroom is the first thing to the woman of the house mentions when she is talking plans.[63]

According to such magazines, cleanliness was to be 'rigidly maintained', and glazed tiles of plain green or blue or mottled colours were beginning to supersede white as symbols of 'freshness and sanitation'. Built-in porcelain toothbrush racks and soap holders were 'the last word in bathroom fittings'. Too crude for interwar readers, toilets were rarely referred to in articles or shown in photographs but did feature in advertisements for 'sanitaryware' showrooms.[64]

The suburban home was undergoing a reorientation. New home designs, including the California bungalow, were heavily influenced by the gospel of scientific management. Efficiency in the use of domestic space was now prized, not least as a way to curb the rising costs of home building.[65] As sociologist Kerreen Reiger concludes, the demise of servants in the early twentieth century 'necessitated a more functional arrangement of working areas for the middle class housewife'.[66] The shrinking kitchen space allowed room for one worker only, thus consolidating the sexual division of labour in the suburban household.[67] Yet, contemporary accounts indicate that designers were not well acquainted with the vagaries of domestic work or household maintenance.[68] 'Women's labors are trebled', wrote Hilda Vane in *Lone Hand* in late 1920, 'and their years shortened by the silly impracticability of the average architect who puts in sinks and washtubs at a height that is

[63] *Western Homes*, October 1929, 48.

[64] *Western Homes*, November 1929. See also *Turner's Western Vogue*, 2, 3 March 1938.

[65] Boyd, *Australia's Home*, 76.

[66] Reiger, *The Disenchantment of the Home*, p. 52.

[67] Lake, 'Historical homes', 52.

[68] Lake, 'Historical homes'.

bound to give the worker a back-ache..., and drains are designed to require the constant attention of the elusive plumber'.[69]

Regardless of form and style, there was little variation in the floorplans of what Boyd called 'the small house' – layouts of five to eight rooms, with separate kitchens and bathrooms and open floorplans without dark hallways, were now standard.[70] Although indoor plumbing had become commonplace around World War I, some houses remained without these conveniences in the 1940s. A resident of working-class Bassendean recalled of the late 1920s:

> Clothes were boiled in the copper over an open fire in the backyard shielded by a piece of corrugated iron. The kitchen in our house was originally unlined, with a wood stove and no running water. We washed up in a basin on the kitchen table. To have a bath before the acquisition of chip water heaters, saucepans of boiling water were poured into a big, concrete tub which was placed on the floor in front of the kitchen stove. The bath was a concrete one which absorbed the small volume of hot water very rapidly. ... The toilet was a little building right at the back of the block. This was common practice in those days.[71]

Running hot water was a preserve of the wealthy who could afford gas bath heaters; chip bath heaters, in which water was heated over a wood fire, were more commonplace in Perth's suburbs.[72] Either way, the hot water supplied from chip or gas heaters 'was rather less than instant or copious', as the heater had to be lit, with the consumer then having to wait fifteen to twenty minutes for the water to be heated.[73] The level of water service provision inside the home mattered deeply, as water provided the means by which settler Australians could uphold and perform the ideals of suburban respectability, those disciplines of behaviour such as cleanliness, sobriety, and thrift. For those of more limited means, maintaining respectability meant living 'with a perpetual sense of threat', as historian Janet McCalman argues, such were its implications for 'the integrity of the personality and the family, the good regard of the neighbourhood and employers'.[74] Even rudimentary water technologies helped housewives, in a context of declining availability of domestic servants, to conform to middle-class expectations of the domestic sphere.

69 Hilda Vane, 'Woman To-Day', *Lone Hand*, 1920.
70 Boyd, *Australia's Home*, 61–62.
71 Cyril Carter (b. 1911, remembering childhood), Oral History, 2007, Bassendean Library, cited in Leigh, 'Bassendean', 352.
72 Barton, 'Household technology'.
73 Davison, 'Down the gurgler', 48.
74 McCalman, *Struggletown*, 19–20.

Despite improvements in household plumbing and the availability of piped scheme water, the domestic duties of most housewives remained physically arduous. It was not only housewives undertaking the work. Reflecting on her childhood in working-class Bassendean prior to World War I, a former resident recalled helping her mother:

> When we came home from school we all used to have to work. We would cart water from the bottom well for the washing tub, we had no water laid on. . . . We would have to cart it up in tubs and kerosene tins to fill the tank so Mum could do her washing.[75]

In more affluent Claremont, a resident recalled Aboriginal women visiting his childhood home on Bay View Terrace to wash laundry on Monday, while in Swanbourne, Aboriginal women also washed and ironed clothes.[76] Other households in the area could send out laundry commercially: Wing Hai's laundry continued to operate nearby into World War II.[77] Interactions with water in early twentieth-century Australian homes were deeply shaped by gender, class, and race.

Detached suburban bungalows required larger lot sizes than cottages and terrace houses, and their lawns and gardens sustained a high demand for water usage. Before World War I, the standard lot in many older suburbs of Perth had a frontage of 13.4 m (44 feet), wide enough for houses with rooms on either side of a hallway.[78] The Californian bungalow and other villa styles reduced housing densities in interwar suburbs because they sat broadly across lots rather than running down them, with frontages wide enough to allow side access for cars.[79] Lots were also large enough for houses to be set back, with ample space for lawns and gardens in the front. Streets were wide enough for grass verges (elsewhere called 'nature strips'), planted with native or imported trees and shrubs, to separate footpath and street.

By the late 1920s, demonstration effects of consumer fashion and the increasing influence of American culture brought elements of a distinct southern Californian townscape to Australian suburbs and country

[75] Winnie Jose (née Best), interviewed by Jennie Carter, 1982.

[76] Bolton and Gregory, *Claremont*, 119; Bropho, *Fringedweller*, 8.

[77] Bolton and Gregory, *Claremont*, 177.

[78] Kelly, 'The development of housing', 245.

[79] Frost and Dingle, 'Sustaining suburbia', 31–32; Kelly, 'The development of housing', 245. Even modest designs were 11.6 m (38 feet) wide, which necessitated square rather than narrow lots, with a 15.2-m (50-foot) frontage to allow side access. As cars became more common, frontages were widened to 16.8 or 18.3 m (55 or 60 feet), comparable to those of parts of West Perth and Claremont, subdivided at the turn of the twentieth century.

towns: palm-planted main streets and public and commercial buildings 'reflecting various interpretations of a Mediterranean style – Mission, Spanish Colonial, Italian Renaissance; ordinary house styles shifting from be-towered and gabled Victorian to the horizontal lines, wide eaves, and veranda fronts of the "bungalow" embowered in an array of exotic trees, shrubs, and flowers'.[80] Australian town planners who visited California's Pasadena – the 'bungalow city' – saw a townscape of detached housing, lush gardens, and dustless streets that was physically and morally superior to that of flats and slums.[81] For both town planners and householders who shared that vision, water supply and sewerage was needed not only for domestic functions but also to allow living on large lots, with manicured green spaces.

SUBSIDISED HOUSING

For many Australian families, the Californian bungalow and its promise of sanitary modernity was merely a dream. In the early twentieth century, State governments became increasingly concerned by the high cost and poor standard of housing available for working-class families. This reflected overseas trends, particularly the emergence in Britain of the Garden City movement, the growth of town planning, and the development of model homes.[82] As they grappled with housing shortages and responded to reports of slum housing conditions, particularly in inner suburbs of Sydney and Melbourne, the idea that homeownership would encourage thrift and create responsible citizens gained bipartisan political support.[83] Conservative and Labor governments intervened in housing markets to facilitate homeownership and provide low-cost housing to low-income earners. These interventions provide evidence of ongoing aspirations to improve the sanitary condition of the 'respectable' working class and served to embed expectations of water provision and disposal that would facilitate the development of new water cultures among a significant part of the population.[84] In New South Wales alone, housing assistance schemes funded 107,000 houses between 1909 and 1943 – equivalent to Adelaide or Brisbane's entire housing stock in 1911.[85]

80 Meinig, *The Shaping of America*, 64.
81 Freestone and James, 'Learning from LA', 49.
82 Booth and Huxley, '1909 and all that'.
83 See for example Davison, *The Rise and Fall of Marvellous Melbourne*, 178.
84 Butlin, Barnard, and Pincus, *Government and Capitalism*, 227.
85 Butlin, Barnard, and Pincus, *Government and Capitalism*, 229.

A total of 38,000 subsidised dwellings were built in Adelaide, Brisbane, and Perth during the same period.[86] In Melbourne, more than 20,000 houses were built between 1921 and 1933 using loans from the State Savings Bank of Victoria (SSBV), 37 per cent of total house construction.[87]

The Queensland *Worker's Dwellings Act* of 1909 was followed by similar legislation in the four other mainland states by 1913. These provided loans for 'persons of limited means' using the mechanism of a state bank or semi-government board that was permitted to raise funds by debentures. Queensland workers with a maximum annual income of £200 who owned a block of land were eligible for a loan for an approved house to a maximum value of £300, but most were unable to meet the eligibility criteria of owning a site and being able to pay a cash deposit of 33 per cent of the value. The Scaddan Labor government established the Western Australia's Workers' Homes Board (WHB) in 1912, which provided freehold or leasehold housing for manual workers, with plans prepared by government architects.[88] In some cities, freehold applicants chose their own sites, so houses were distributed across the metropolitan areas, rather than clustered in planned subdivisions. In Queensland, responsibility for Workers Dwellings was transferred to the State Savings Bank of Queensland (SSBQ) in 1916. Under amendments in 1919 and 1923, the maximum loan was increased and the required deposit reduced, and a parallel scheme provided perpetual-lease housing on government-owned land, as in Perth.[89] People of 'small means' were not always the beneficiaries. In Adelaide, the Advances for Homes scheme of 1910 attracted successful applicants who were skilled manual, clerical, and professional workers.[90]

New South Wales and Victoria's earliest housing acts addressed the problem of slum housing by empowering municipal councils in Sydney and Melbourne to borrow funds to provide housing, following British practices. Until 1890, when the Victorian government legislated minimum street widths in new subdivisions, working-class housing was constructed without any regulation.[91] Municipal councils were authorised to inspect

86 Kilner, 'The evolution of South Australian urban housing policy', 272; Rechner, 'The Queensland workers dwelling', 265.
87 Ward, 'The development of Melbourne in the interwar years', 332.
88 Menck, 'A thematic history of Government housing in Western Australia', 40.
89 Rechner, 'The Queensland workers dwelling', 266–71.
90 Kilner, 'The evolution of South Australian urban housing policy', 263.
91 Barrett, *The Inner Suburbs*; Lemon, *The Northcote Side of the River*, 93–94. Municipal councils were not required to set minimum lot sizes until 1921.

and condemn insanitary housing under the *Public Health Act* of 1883; the New South Wales government granted the City of Sydney power to resume slums in 1905 and set up a Royal Commission into the Improvement of the City and Suburbs in 1908. By 1912, slum clearance in Sydney had displaced around 7000 people.[92] After the 1913–17 Royal Commission into Housing Conditions in Melbourne, the Conservative Lawson government passed the *Housing & Reclamation Act* of 1920, extending the power of the SSBV to acquire land and create a credit foncier department (that facilitated loans where principal and interest were repayable over extended periods). The SSBV included an architectural branch, developing a range of 'approved' housing styles, mostly based on Californian bungalows.

Garden city ideas were embodied in planned residential suburbs for those of 'limited means' in Sydney (Dacey Garden Suburb, or Daceyville, and Matraville), Adelaide (Colonel Light Gardens), and Melbourne (Garden City). Developed on sandy and swampy Crown Land under the *Housing Act* of 1912 by the McGowen Labor government, Daceyville was by 1919 lauded for 'the beauty of the gardens and lawns, the practical comforts of the interiors, [which] give an impression of a different civilisation from that of the crowded areas of the city' (Fig. 5.3).[93] The plans included wide streets, open spaces for parks, and land reserved for shops and a school. After World War I, fifty cottages were made available for incapacitated soldiers and war widows. By 1920, 315 detached and semi-detached houses had been built for rent that was below market rates. Most tenants were working and lower middle class, such as tramway workers and government clerks. Designed as a result of architectural competitions for cheap housing, houses included a verandah, living room, one or two bedrooms, kitchen, and bathroom and pantry at the rear. The kitchen led to a laundry with adjacent lavatory, both under the main roof. Water, sewerage, gas, and electricity were connected before building, so no rear lanes for service access were provided. The Housing Board assumed 'liability for the excess water rates and garden fees' to maintain the 'garden character of the Suburb', such was its emphasis on the upkeep of Daceyville's gardens.[94]

Adelaide's Colonel Light Gardens was the work of government Town Planner Charles Reade, a New Zealand-born journalist who was prominent in Britain's Garden City movement and had lectured across

[92] Fitzgerald, *Sydney 1842–1992*, 224. [93] Cited by Freestone, *Urban Nation*, 178.
[94] Cited in Hoskins, 'Constructing time and space', 8.

FIGURE 5.3. Daceyville street, housing, and road under construction, 1914 (photographic album, Housing Board of New South Wales, NSW State Archives).

Australia in 1914–15. The Vaughan Labor government established a state planning department with power to vet suburban plans and design new communities to ensure 'proper sanitary conditions', including adequate drainage and, where economical, sewerage.[95] Reade's 1917 plan set out nearly 600 lots with ample public space. Early development was slow, and in 1924, the South Australian government's Thousand Homes Scheme, specifying that homes had to be built within six miles of the City of Adelaide and grouped in planned areas to reduce costs, targeted Colonel Light Gardens to help address housing shortages.[96]

At Colonel Light Gardens, workers could borrow up to £700 from the State Bank, with terms of forty-two years for stone or brick houses and twenty years for wood and iron. Ten floor plans were available for houses of four- to six-room homes, with a choice of fourteen facades. Utility rooms (kitchen, bathroom, toilet, and laundry) were all under the main roof, and electricity, gas, water, and flush toilet were provided. Detailed

[95] Freestone, 'An Imperial aspect', 168–70. [96] Freestone, *Urban Nation*, 178–79.

by-laws were set out under the *Garden City Act* of 1919, administered by a Garden Suburb Commissioner. Bathrooms had to include 'a bath not less than 5ft in length'; corrugated iron rainwater tanks, set sufficiently high on a stand that the bottom was on the same level as the wash trough and taps, were mandatory.[97]

In Victoria, investigations into metropolitan housing conditions prompted the creation of new institutions to eliminate slums and provide social housing. From 1926, drawing on the example of Britain's State-aided Housing Schemes, the SSBV purchased 18 hectares of land at Fisherman's Bend on the bank of the Yarra near Port Melbourne to develop Garden City. The 1936–37 report of the Housing Investigation and Slum Abolition Board found 3000 Melbourne houses, most dating from the 1860s, to be unfit for habitation, with the city facing a shortage of 25–30,000 low-rent houses.[98] Middle-class commentary on these housing standards invariably extended to critical assessments of the inhabitants and their character; as historian Tony Birch notes of slums in Melbourne's Fitzroy, they were portrayed as 'the antithesis of the decent family home environment'.[99] Resident mothers were especially condemned as immoral and ignorant; their children would only thrive if 'rescued from the sordid surroundings and vicious influence of the slums', argued Melbourne reformer Frederick Oswald Barnett in the early 1930s.[100] The Housing Commission of Victoria was established under the *Housing Act* of 1937 to address the problem, although the social stigma of the environmentally determinist reformist campaigns lingered into the 1960s.[101]

Restrictions in funding slowed the construction of workers' homes during the Great Depression. In Perth, at least one-third of workers' home residents struggled to make repayments.[102] A philanthropic organisation, the McNess Housing Trust, built 255 houses in cooperation with the WHB. The standard plan was an 1184 square foot (110 square metre), four-room timber cottage. Eligible applicants included aged pensioners (mostly women), invalid pensioners with dependent families, widows with young children, and deserted wives aged over 21 with dependent

97 Thousand Homes Scheme, Colonel Light Gardens Specifications and By-laws under the Garden Suburb Act, 1919, no.11, *South Australian Government Gazette*, 29 September 1921, 719; Garnaut, *Colonel Light Gardens*.

98 Housing Investigation and Slum Abolition Board, *First (Progress) Report*.

99 Birch, 'These children have been born in an abyss', 8.

100 Barnett, 'The economics of the slums', 48.

101 Mayne, 'A just war'.

102 Menck, 'A thematic history of Government housing in Western Australia', 54.

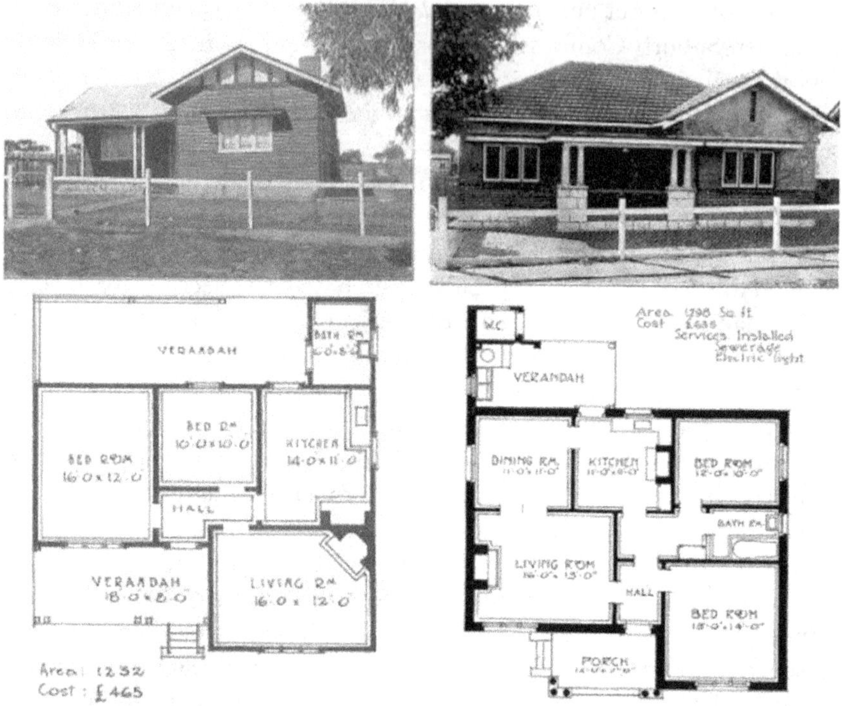

FIGURE 5.4. Least and most expensive homes (WHB Annual Report 1934).

children and 'of good character'.[103] As building activity resumed in the late 1930s, brick WHB houses were built in working-class Mosman Park and Fremantle, close to industrial workplaces, and on the outskirts of Claremont and Nedlands. Plans for one house built in Nedlands in 1939–40 show a 1019-square foot bungalow with a small front porch leading to a hall and a doorway to a lounge and dining room, in turn leading to a kitchen. On the other side of the house were two bedrooms with a bathroom in between. At the rear of the house, adjacent to the kitchen and back verandah, was a combined laundry (with copper and troughs) and bathroom (with shower, but no bath) and a separate toilet. The remainder of the back verandah could be divided to make a sleepout off the second bedroom.[104]

[103] Menck, 'A thematic history of Government housing in Western Australia', 71–77.

[104] Specifications and plan, Type No.103 B, Workers Homes Board Brick House to be erected on Lot 230 Dalkeith Rd, Nedlands for Mr H Rowe, dated 13 October 1939, SROWA, AU WA S442 – cons1613 1939/0476.

A more austere house style developed in Brisbane to reduce costs during the Great Depression, with shorter stumps, a porch instead of verandas, and fibro-asbestos roof tiles and walls.[105] Kitchens now came with built-in cupboards, sinks and taps, and electric and gas appliances. Since 1925, all Queensland Worker's Dwellings designs had included an internal bathroom; now, tiling to bathroom walls around the bath and on the floor, a bath, sink, washbasin, a thousand-gallon (4546-litre) rain-water tank, hot water service, and septic system were included in each house.[106] Tanks were supplied to collect run-off from corrugated iron or asbestos roofs. Instead of corrugated iron walls and timber floors, bathrooms began to be finished in hygienic, easy to clean materials, such as glazed tiles, art metal, or fibro-cement on walls, and with terrazzo or tiles on the floor. Gardens were developed slowly on cleared building sites, in a climate suited to jacaranda shade trees, as well as frangipani, paw paws, and mangoes. As much of Brisbane remained unsewered, sewage was a public health hazard, discharged into local watercourses through drains.[107] The 1935 State Advances Corporation handbook noted that where drains for household grey water were not available and the soil was not absorbent, the 'installation of absorption trenches or sumps to receive house drainage [was] strongly condemned'. It was thought better to 'broadcast the water' across the lot. Estimated costs ranged from £38 (equivalent to eleven weeks of wages for an unskilled worker) for installation of a septic system, £30 to connect to sewerage where it was available, and £12 for drainage into the water table.[108]

House plans published in booklets for Adelaide's Thousand Homes scheme (in 1924), Sydney's GSBNSW (1929), Melbourne's SSBV (1929 and 1941), Perth's WHB Annual Reports (1933), and Brisbane's SSBQ (Figure 5.5) show the changes taking place in subsidised housing during the interwar period. They are also indicative of the evolution of new suburban housing that was built by the private market, without subsidy. To assist applicants in choosing a plan, the price of each house, size in square feet, and information about electricity, gas, and sewerage were

105 Rechner, 'Houses for Queenslanders of small means?' 58, 66, 146, 189–90.

106 State Advances Corporation of Queensland, 'Designs of dwellings', January 1935, https://nla.gov.au/nla.obj-52796600/view?partId=nla.obj-94988544#page/n41/mode/1up, accessed 28 March 2021.

107 Greenwood, *Brisbane 1859–1959*, 406.

108 State Advances Corporation of Queensland, *Designs of Dwellings* (Brisbane: Government Printer, 1935). The Basic Wage in 1935 was £3 8 shillings per week. Vamplew, *Australians*, 157.

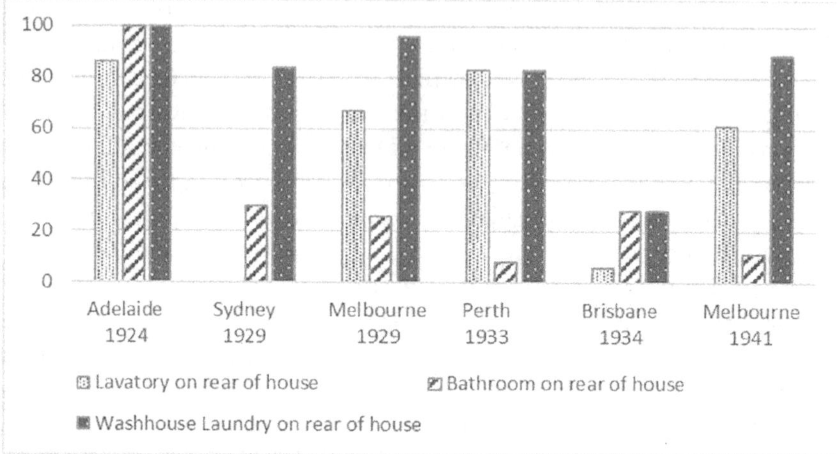

FIGURE 5.5. Proportion of house plans under Australian subsidised housing schemes with lavatory, bathroom, or laundry on the rear of the house, 1924–41. (*Sources:* State Savings Bank of South Australia, Thousand Homes Scheme Plans, 1924, kindly supplied by Philip Knight, Colonel Light Gardens Historical Society; State Savings Bank of Victoria, Design Book: Brick Dwellings, March 1929, https://nla.gov.au/nla.obj-52850318/view?partId=nla.obj-101549704#page/n0/mode/1up; State Savings Bank of Victoria, Design Book: Housing Loan Conditions, March 1941, https://nla.gov.au/nla.obj-528568 58/view?partId=nla.obj-102817791#page/n1/mode/1up; Government Savings Bank of New South Wales, Advances for Homes Department Plans, January 1929, https://nla.gov.au/nla.obj-436364 36/view?partId=nla.obj-43636545#page/n1/mode/1up, accessed 29 March 2021; Workers Home Board, Annual Report, 1933; State Advances Corporation of Queensland, Designs of Dwellings, January 1935, https://nla.gov.au/nla.obj-52796600/view?partId=nla.obj-94988544#page/n41/mode/1up, accessed 28 March 2021).
Note: Samples: Adelaide n = 14; Sydney n = 49; Melbourne (1929) n = 27; Perth n = 12; Brisbane n = 32; Melbourne (1941) n = 28.

provided. None of the plans are identical in architectural style or layout, but the provision of water, sewerage, and septic tanks reflects changing expectations in the interwar period, as water for cleaning, cooking, and bathing shifted inside the home, along with toilets.

These shifts over time can be quantified via the housing plans for these subsidised schemes, which reveal the unevenness of the sanitary transition across the five capital cities. Of the Adelaide plans in the mid-1920s, 86 per cent show a lavatory attached to the rear of the house, under the roofline, with 14 per cent of these in an internal bathroom. In Sydney, five years later, no workers' home plans feature an inside lavatory; all would have had backyard lavatories (or 'dunnies'). In Melbourne, two-thirds of plans include a lavatory on the rear of the home; in 1941, this figure has dropped to 61 per cent, by which time only 7 per cent of plans show an outside lavatory. None of the 1929 Melbourne plans indicate a toilet inside a bathroom, but this feature appears in nearly a third of the 1941 plans. Lavatories on the back verandah appear in 83 per cent of Perth's WHB house plans in 1933. In 91 per cent of Brisbane plans in 1934, there is a dunny in the backyard or under the house, made possible by the elevated Queenslander style in the subtropical city. Almost all plans show a wash-house at the rear of the house, often next to a separate lavatory; in Brisbane, the wash-house is outdoors in nearly three-quarters of plans. By 1941, 7 per cent of Melbourne plans show an internal laundry. As this data suggests, the placement of bathrooms inside the house, convenient to the bedrooms, began to occur at the end of the 1920s in Sydney and Melbourne. While subsidised dwellings were simpler than those built by the private sector, by World War II they all came with running water and most had internal bathrooms. Although it appeared the transition to sanitary convenience was complete, much of the suburban population was still living in older housing stock, or even makeshift dwellings, where the upholding of standards of respectability was hard won.

LEFT BEHIND

Born in East Perth in 1924, Jesse Grasso grew up in 'a very old house' between Claisebrook Road and the main railway line. Her Italian parents ran a small store in a neighbourhood of semi-detached houses, with a foundry opposite and an oil company, scrap metal yard, biscuit factory, and gasworks operating nearby. Traditionally a site for noxious trades, East Perth was designated 'light industrial' under the *Town Planning Act* of 1928, with future industrial development directed to outer suburbs such as Bassendean, Maylands, and Welshpool. New warehouses and factories were less noisy and smelly than previous land use, but the area retained its working-class character. East Perth was typical of inner industrial districts in the five cities, the population of which was in decline in the interwar years. Between 1921 and 1947, the population of the City of Sydney fell by almost 8 per cent; that of the City of Melbourne and five adjoining municipalities fell by 6 per cent.[109] Once available land in these

[109] Fitzgerald, *Sydney 1842–1992*, 225; Census of the Commonwealth of Australia (Population).

areas was taken up, expansion of industrial and commercial buildings resulted in demolition of parts of the housing stock and higher residential densities. Out-migration of skilled workers from fully developed areas drove the growth of new suburbs, but people who remained lived in rundown housing.[110] 'I think we were all battlers in East Perth at that particular time', Grasso recalled, 'because it seemed like only the poorest people lived in East Perth and people used to think that too'.[111]

The outbreak of war added to the pressures on Australian cities. In Melbourne alone, the city's population increased by over 150,000 between 1939 and 1945, as workers moved to the city to take up jobs created by the War. The Commonwealth imposed building restrictions that fixed the number of dwellings, pegged rents at 1939 levels, and prohibited the eviction of tenants.[112] Australian and American military personnel added to the demand for housing; some dwellings were shared between two or more households. At the start of the War, Melbourne's housing shortfall was estimated at 25,000. The nation's housing deficit was estimated at 120,000 dwellings, and as high as 350,000 by 1945.[113] After the 1941 census was postponed indefinitely because of the War, the Commonwealth Department of Post-War Reconstruction, the University of Melbourne, and a group of local businessmen funded a social survey under the direction of Wilfred Prest, an economist at the University of Melbourne, to obtain information about housing conditions to inform future policy. Almost 8000 households from across the metropolitan area were surveyed over a seventeen-month period, with interviewers recording details such as the type of dwelling, the condition of the garden and the extent of cultivated space, the location of the water closet, the presence of a sink and water tap in the kitchen, and whether the kitchen was 'ever used as a Bedroom'.[114] The investigators encountered a broad spectrum of living conditions, such as homes that were 'well-built, attractive ... well-kept garden', while others were 'in bad state ... bathroom disgraceful. Kitchen and bathroom horribly untidy and dirty'.[115] In inner Melbourne, where more than two-thirds of housing was rented, many

110 Neutze, *Urban Development in Australia*, 65–70; Kelly, 'Picturesque and pestilential'.
111 Cited by Christensen, 'Industrial East Perth in growth and decline', 70. See also Christensen, 'East Perth 1916–1981', 24–25; Gregory, *City of Light*, 162.
112 Troy, *Accommodating Australians*, 38; Darian-Smith, *On the Home Front*, 86.
113 Darian-Smith, *On the Home Front*, 84.
114 Australian Data Archive. https://dataverse.ada.edu.au/file.xhtml?fileId=8787&version=3.0, accessed 15 June 2021.
115 Cited in Davison and Lack, 'Planning the new social order', 41.

householders interviewed for the Prest Survey indicated a desire to move, but better quality accommodation was unavailable.

The Prest Survey revealed disparities in domestic water facilities throughout Melbourne. In newer suburbs, internal bathrooms, laundries, and hot water services were included in bungalows and villas at the outset; in nineteenth-century houses, they were later additions. In inner-city Collingwood, 60 per cent of households in the sample obtained water from an outside tap, compared to 5 per cent in middle-class Camberwell. Ninety-five per cent of houses in Collingwood and working-class Northcote had no hot water service and had to heat water for baths and clothes washing on a stove; in Camberwell, the figure was 59 per cent.[116] An investigator noted of a South Melbourne housewife that she 'had done her best to give her daughter a good start, but she wanted a house with a decent bathroom [that in the present house being outside in the wash-house]. She thought it would be a good thing if the daughter could have a bath every day – like the girls at work'.[117] The Melbourne inner-city bathroom was usually located on a verandah or attached to a cramped wash-house in the backyard. Sewerage diverted human waste from local drains and creeks, but low-lying areas remained poorly drained and prone to flooding, due to incomplete forming of streets and gutters.[118] Toilets continued to be located at the bottom of backyards.

Many Aboriginal people had been forced onto small reserves that were often situated near rubbish dumps and sanitary depots, where there were inadequate water supplies and sanitation facilities.[119] Robert Bropho recalled that in the early 1940s at the Eden Hill camp on the eastern outskirts of Perth, the 'only water supply was from the local tip on the hill near the swamp' or from shallow wells they dug in the ground.[120] In their tin billies, they collected water with 'tadpoles and the slime on the top', 'with a bit of flavour in it from human piss and human shit'.[121] By this time, most members of the Aboriginal community at Salt Pan Creek in Sydney had been forced to move to La Perouse, after complaints from local white residents that the camp be removed due to poor hygiene and noisy disturbances.[122] Near Perth's affluent Claremont, the council

116 Darian-Smith, *On the Home Front*, 97.
117 Cited in Davison and Lack, 'Planning the new social order', 42.
118 Lemon, *The Northcote Side of the River*, 190–91.
119 Haebich, *For Their Own Good*, 234. 120 Bropho, *Fringedweller*, 27.
121 Bropho, *Fringedweller*, 30. 122 Goodall and Cadzow, 'Salt Pan Creek', 204.

FIGURE 5.6. A woman washing clothes in the backyard of a Collingwood house, c. 1935 (State Library of Victoria).

bulldozed the Galbamaanup camp in the early 1950s.[123] Of the council's decision, the *West Australian* reported, 'the Nedlands road district was not a fit and proper place for a native camp [because] the white population was spreading in the Swanbourne area and ... the board could not allow a "black camp in a white area"'.[124]

At the end of World War II, most non-Indigenous urban Australians enjoyed domestic water facilities that, in an overall sense, compared favourably to those of the USA and Britain. In 1947, between 87 and 92 per cent of all houses in the five Australian cities had a bathroom, while only 77 per cent of households in American cities enjoyed exclusive use of a bathtub or shower in 1940.[125] Bathrooms were by no means typical in British working-class homes – in most cases the kitchen sink

[123] Letters and Memos, 9–20 December 1912, 1 March–1 May 1913, 'Aboriginal Camping Ground Claremont', File 1644/12, Aboriginal Planning Authority, Acc 652, AN 1/3, J.S. Battye Library.

[124] *West Australian*, 4 April 1951, cited in Cook, *That Was My Home*, 155.

[125] Census of the Commonwealth of Australia, *Dwelling Summary Bulletins* (1947); Gordon, *The Rise and Fall of American Growth*, 120.

FIGURE 5.7. Fitzroy houses, with backyard dunnies backing on to a lane way and washing on the line, c. 1935 (State Library of Victoria).

was used for daily washing, with a tin tub used for a weekly bath.[126] In Adelaide, 83 per cent of dwellings were connected to sewerage or a septic tank, exactly the same figure as in US cities (in Melbourne and Sydney, the figures were 82 and 78 per cent).[127] These aggregates tend to mask the situation in disadvantaged areas of the five Australian cities. In Britain, one in four working-class households had piped hot water in 1942, more than their Melbourne counterparts in 1947.[128] As in Britain, Australian working-class households were likely to have an external laundry or wash-house, with small yards that offered little space for drying clothes (Figure 5.7). Despite piped water and sewerage, not all Australian inner suburban bathrooms and toilets were internally located, necessitating compromises in privacy and comfort.

The unevenness of the household sanitary transition continued after World War II. The Commonwealth Housing Commission reported in 1944 that 'electric light and power, more than anything else, hold out the hope of help to the housewife, as well as the promise of more

[126] Daunton, *House and Home in the Victorian City*, 247.
[127] Census of the Commonwealth of Australia, *Dwelling Summary Bulletins* (1947).
[128] Daunton, *House and Home in the Victorian City*, 242.

attractive and comfortable homes'.[129] As historian Tony Dingle notes, 'there was a rising public expectation that everyone would have better housing in the new post-war world, and specifically that every family deserved to occupy and own its own detached home.'[130] Over a decade later, however, a survey found these standards to be still out of reach of many Australian households. The proportion of households in Sydney, Melbourne, and Brisbane owning an electric washing machine was only 39 per cent, and only 26 per cent of homes had running hot water in the laundry, compared to nearly 50 per cent in the kitchen.[131] While washing machines were unaffordable for many households into the mid-1950s, such was the desire for cheap and easy ways of achieving cleanliness that new washing powder products, promoted through advertising that targeted housewives, sold strongly.[132] It was only as wartime austerity slowly gave way to post-war prosperity that suburban Australians could fulfil those 'stoked-up dreams of the depression and war generations', as Barbara and Graeme Davison observe. 'Behind their dreams of domestic privacy lay the memory of over-crowding; behind their craving for security lay the shadow of eviction.'[133]

[129] Commonwealth Housing Commission, *Final Report*, 119.
[130] Dingle, 'Necessity the mother of invention'.
[131] Webber, 'Romancing the machine', 175.
[132] Blackburn, 'The "consumer's" ethic'.
[133] Davison and Davison, 'Suburban pioneers', 43.

6

Watering Suburbia

'Shocking, disgusting', recalled a middle-class woman who, in the post-World War II period, made a home in the suburb of Mordialloc, 24 km from the centre of Melbourne. 'There was no sewerage, no septic tanks, no roads; it was unbelievably bad. I wish I had photos of it.' Located at the edge of the Carrum Swamp, bayside Mordialloc was a typical post-war suburb, in which house construction outran the provision of sewerage and sealed roads.

> I felt sorry for the people who lived further down. We were lucky because we were just in from the main road, which was a little bit better. They used to come up the street here, especially in the winter time, and they'd have to have another pair of shoes in a bag so that they can wipe the mud off their feet before they could go anywhere . . . you couldn't just take the baby for a walk. The pram would be stuck in the mud and the rocks.[1]

Within two years of the outbreak of World War II, pundits and political parties began to consider and plan for peace. In 1941, before Japan entered the war, Ian Clunies Ross, Professor of Veterinary Science at the University of Sydney (and, later, head of the Commonwealth Scientific and Industrial Research Organisation) and a regular radio commentator on the ABC, wrote a booklet, *Should We Plan for Peace?* Clunies Ross was troubled by the 'Australia will do me' approach where people took 'an adequate pride in the Harbour Bridge, the St. Kilda Road, the Brisbane Town Hall, or the view from King's Park in Perth', while issues of housing provision, low population growth, new industries, and

[1] Cited by Murphy and Probert, 'Anything for the house', 280.

the quality of education were unresolved. Clunies Ross denounced 'slums', houses without 'baths or stoves, families living in one room', arguing that housing should be regarded 'not as a commercial undertaking but as a social responsibility' for which the Commonwealth government should take the lead. Once funds were available, Clunies Ross posed the question of whether new housing might be 'in great flat buildings ... putting our people up in the air instead of spread out in their little slums'. Having travelled widely in Europe, the United States, and Japan, he favoured planned parks, playgrounds, and gardens. His fear about plumping for 'the Australian type of individual house and grounds' was that it would exacerbate 'the haphazard scattering of unsightly jerry-built erections which have spread like foul tentacles over Sydney's green belt'.[2]

In 1941, the Commonwealth Labor government created a Department of Post-War Reconstruction, with Treasurer Ben Chifley as the Minister.[3] The new department was applauded by journalist C. E. W. Bean, a household name thanks to his multi-volume official history of Australia in the war of 1914–18, who had in 1907 dubbed Australia's inner cities as 'slums without poverty'.[4] In his best-selling *War Aims of a Plain Australian* (1943), Bean decried the lack of strong town planning bodies in Sydney and Melbourne, pointing out that attempts at metropolitan government had been defeated by local councils and landowners in both those cities, while Brisbane had overcome such obstacles to create city-wide government. Bean argued that roads, railways, water supply, sewerage and drainage, electricity, schools, and playgrounds should all be overseen by a single authority. An advocate of inner-city slum demolition and the health benefits of a suburban garden, Bean wanted suburban subdividers to allocate 10 per cent of their land to parks and recreation, and where that was not possible, they should provide 'the equivalent in money'.[5]

House building virtually ceased during World War II, with both materials and labour directed to the war effort. By the time the nation's 900,000 armed services personnel returned to civilian life, they and their families, and much of the rest of the community, faced a severe housing

[2] Clunies Ross, *Should We Plan*, 11, 16, 42, 47.
[3] Macintyre, *Australia's Boldest Experiment*.
[4] Charles E.W. Bean, 'The Australian city', *Sydney Morning Herald*, 8 June 1907.
[5] Bean, *War Aims*, 124–27.

shortage, estimated in 1945 as 480,000 dwellings nationwide. The 1947 census recorded 877 dwellings for every thousand households.[6]

The Commonwealth State Housing Agreement of 1945 provided loan funds to the states to provide rental housing for low-income families but also to provide homes to ex-service personnel through the Commonwealth War Services Homes Commission. Service personnel could also take out forty-five-year loans at less than 4 per cent interest, well below the prevailing market interest rate, with only a modest deposit required.[7] Asbestos sheeting ('fibro-cement sheeting' or 'fibro'), cheap, light, and easy to cut, was substituted for scarce bricks and weatherboard. Typical housing commission homes in cities and country towns might have a brick or timber portico and sometimes external walls of brick and timber but invariably fibro lining inside. Hundreds of thousands of backyard dunnies in every growing suburb were built of fibro, with tin roofs.

Robert Menzies led a Liberal–Country Party coalition to victory in 1949, promising an immediate end to restrictions on petrol, car purchases, and building materials. Long an advocate of 'one little piece of earth with a house and a garden', homeownership was central to Menzies' liberal vision for Australia; his government introduced incentives that raised the share of owner-occupied housing from 53 per cent in 1947 to 70 per cent in 1961, compared to 41 per cent in the UK in 1961.[8] The conservative government, who had always favoured home ownership above public rental housing, amended the Commonwealth State Housing Agreement in 1954 to allow the sale of houses to sitting tenants, but only 6 per cent had taken up the offer by 1956. Housing Commissions also built high-rise blocks of flats, notably in inner Melbourne, where giant structures replaced thousands of terrace houses, demolished as 'slums'. Sydney saw fewer tower blocks but more three-story blocks of flats.[9]

In middle ring suburbs, the State Housing Commissions had to compete with private development companies to acquire sites suitable for new house building. Many of these sites, especially in Sydney and Melbourne,

[6] Frost, 'Town and country in the Menzies years', 237; Davison and Dingle, 'Introduction', 9.

[7] Hill, *Housing Finance in Australia*, 46–47.

[8] Frost, 'Town and country in the Menzies years', 234; Menzies, 'The forgotten people (1942)'.

[9] For a detailed account of the Commonwealth State Housing Agreement, see Troy, *Accommodating Australians*.

had been subdivided in the 1920s and 1930s, following the electrified suburban rail networks. Reticulated water always came first, as local councils would rarely permit the sale of housing blocks without water, but connection to the sewer could wait, as every city continued to operate a night soil collection system. Household collection was usually a municipal responsibility, paid for out of council rates, while the final disposal point, whether sewage farm or dumping in a river, bay, harbour, or the ocean, came under the auspices of statutory water boards or their equivalent. At the 1947 national census, 200,000 dwellings across all five mainland capitals still had a pan service, 21 per cent of the total number of dwellings. Brisbane had the highest proportion, 55 per cent, and Adelaide the lowest, 17 per cent. The association of such unsewered areas with outbreaks of poliomyelitis failed to hasten the expansion of suburban sewer networks, which continued to lag long after a mass vaccination program commenced in 1956.[10]

Housing shortages and restrictions on their mobility forced many Aboriginal families to erect 'illegal' camps in the outskirts of cities. Sydney had a plethora of such camps, the largest at La Perouse, on the northern headland of Botany Bay. By 1948, up to 300 Aboriginal families were living in what the press described as 'hovels and tents', built of bags, tin, and brushwood.[11] They had no sanitary facilities, no electricity, no garbage collection, and only one communal water tap. The Aborigines Welfare Board wanted to demolish the shanties and offered the occupants replacement tents. The residents rejected the offer. Mr H. J. Simms, who had lived all his life on the settlement, reportedly said to the Sydney *Tribune*, 'my ancestors lived on this land before there was such a thing as an Aborigines Welfare Board'.[12]

The housing shortage continued to be so severe in the early 1950s that there were reports of Aboriginal people selling their illegal huts to 'New Australians' – migrants and refugees from Poland, Russia, Ukraine, and Malta – for £90.[13] While immigrants could often find employment in local hospitals or the new oil refinery, their housing options were limited in the early post-war years to overcrowded migrant camps, in army-style

[10] Census of the Commonwealth of Australia, Dwelling Summary Bulletins. See Southcott, Crosby, and Stenhouse, 'Studies on the epidemiology of the 1947–1948 epidemic of poliomyelitis in South Australia'; Lewis, *The People's Health*.

[11] *Sun* (Sydney), 1 September 1948; *Canberra Times*, 16 October 1951.

[12] *Tribune* (Sydney), 2 October 1948.

[13] *Newcastle Morning Herald*, 16 October 1951.

FIGURE 6.1. New development in the Brisbane suburb of Inala, 1962 (courtesy of Queensland State Archives, ITM1541665, Housing Commission Estate, Inala, Brisbane).

tin sheds. Migrants, Aboriginal people, and Australian families who could not afford to buy a house were the main tenants of State housing commission developments. The entire suburb of Inala on the fringes of Brisbane, constructed in the early 1960s, housed a population of 16,000 people (Figure 6.1).[14]

In an effort to improve Australian housing standards, the Commonwealth government provided advice on house design and garden layout, based on scientific principles. Architect Walter Bunning's *Homes in the Sun* (1945) emphasised the importance of orientation, cross ventilation, roof overhangs for shade, and large windows for natural light. Architects' plans were popularised in the *Australian Women's Weekly*, to be found in almost every Australian household, and in other popular magazines, including the monthly *Australian House and Garden*. Owner builders, who oversaw the construction of up to half of all houses on suburban blocks, consumed these magazines and swapped well-thumbed copies of house design booklets. In 1952, the magazine ran a series of articles under the heading 'a house you can build yourself' with 'smartly

[14] Brisbane City Council Annual Report (BCCAR) 1960–61, 34.

FIGURE 6.2. Bulli Street in the Melbourne suburb of Moorabbin, 1960s (Kingston Historical Society).

modern' plans 'to meet the needs of every average, home hungry Australian'.[15]

When Robin Boyd wrote *Australia's Home* in 1952, he worried as to how new European migrants would cope with the post-war reality: 'Out at the end of the Australian suburb, where raw new houses stood in shadeless yellow grass beside the rutted tracks, one seemed to suffer all the disadvantages of city life and enjoy none of the advantages.'[16] Boyd continued this theme in 1962 in *The Walls Around Us*, a book he wrote and illustrated for younger readers:

> Thousands of the proud new homes must sit for a while in a sea of mud because the road in the front is unmade, and many are without sewerage, because the government departments which lay the roads and the pipes cannot keep pace with the growth of the houses.[17]

Figure 6.2, showing a street in the south-eastern Melbourne suburb of Moorabbin, provides an example of this suburban frontier housing.

[15] On the growth of owner occupation, see Holland, 'The comfortable house', 197–217; Holland, *Emoh Ruo;* Greig, *The Stuff Dreams Are Made Of*, 97–120.
[16] Boyd, *Australia's Home*, 277. [17] Boyd, *Walls Around Us*, 66.

THE SUBDIVISION BOOM, THE SEPTIC TANK, AND TACKLING THE SEWERAGE BACKLOG

A subdivision land boom in Sydney and Melbourne in the interwar years had left thousands of unsold lots, which came onto the real estate market in the 1950s. In both cities, these lots were within walking distance – a few kilometres – of electrified railway lines or tramway systems. The rapid growth of car ownership in the 1950s and 1960s meant that subdividers in all cities could go well beyond the public transport spines. Increasingly, lots were sold in large numbers in named marketed estates, including former racecourses on the Bay in Melbourne between Mordialloc and Frankston, with the subdivisions rapidly outstripping sanitary services.

Owner builders were happy to embrace new subdivisions that, by law, offered reticulated water, but lots remained relatively cheap because they were without made roads, guttering, pavements, or sewerage. All the other attributes of suburbia, including stormwater drains, came later, paid for by the householder, usually by a separate council levy. From the mid-1950s, septic tanks, involving an underground tank, were connected to toilets, showers, baths, the laundry, and the kitchen. The grease trap of the tank had to be cleaned regularly, and sometimes tanks had to be flushed out. In the nuclear families that dominated the new suburbs, this task was invariably allocated to the oldest male member of the household. For urbanist Patrick Troy, dependence on an inferior septic system, rather than expenditure on sewerage, 'represented a failure of the centralised system'.[18]

Post-war Brisbane faced a shortage of labour and resources to cope with the demand for better roads, water, electricity, and sewerage.[19] The backlog in sewerage reticulation work emerged as election issues in the 1949 and 1952 Council elections. By 1949, only 4071 Brisbane houses were connected to sewerage.[20] Eagle Farm pumping station, opened in 1955, was designed to cope with 364 million litres daily, but the maximum daily sewage load was 155 million litres in 1955–56.[21] In 1952, *Truth* newspaper described 'Brisbane with its vast unsewered areas, in which noisome outhouses stand in endless ranks, links sentry boxes of the Army of Disease'.[22] Declaring the city the 'most backward of all

[18] Troy, 'Life and times', 15. [19] Cole, *Shaping a City*, 123–24.
[20] Cole, *Shaping a City*, 128.
[21] BCCAR 1954–55, 30; 1955–56, 40; Cole, *Shaping a City*, 158.
[22] 'Sewerage scandal worse', *Truth* (Brisbane), 20 January 1952.

FIGURE 6.3. Mother and daughter in suburban Melbourne 1953 (photographer: Len Drummond).

Australian capital cities', the journalist decried a city of approximately 117,000 dwellings, with only 48,719 sewered, leaving 60 per cent unsewered, 'doomed to the filthy, appalling day-cart system' (Figure 6.4). With hundreds of cases and two deaths from Sonne dysentery in 1949 and 7000 homes under construction each year, the issue demanded

FIGURE 6.4. Outside toilets in the backyards of the unsewered suburb of Norman Park, Brisbane, in the 1950s (Brisbane City Council Archives).

attention.[23] By 1961, only 46 km of sewerage mains and 466 km of reticulation mains had been constructed since World War II, with a total of 67,407 premises sewered, only 43 per cent of the number of properties with water connected.[24]

In Melbourne, the MMBW reported that up until 1945 the system had 'kept pace closely with housing development', with only 11,000 unsewered houses, only 4 per cent in the metropolitan area. Labour and material shortages and the housing boom increased that number to 22,000 by June 1951.[25] By June 1953, 50,000 people lived in unsewered homes.[26] The system, designed in 1892 for one million people, could no longer deal with the increasing sewage flows in a city of 1.35 million. Industrial development 'which could not be visualised by the early designers' caused further strain as manufacturers discharged large volumes of polluted water into the sewerage system (16 per cent of daily flow). This additional load reduced capacity for coping with stormwater, which resulted in the system becoming overloaded in wet weather.[27] In 1951, the Board found it would take approximately £13 million, 2000 men, and ten years to address the backlog, one that was growing every year.[28] Its annual reports hint at despair of ever catching up. With new builds of 10,000 to 12,000 homes per annum, and the Board only able to sewer 7000 each year, the Board reported 'the position will continue to deteriorate'.[29]

23 'Sewerage scandal worse'. 24 Cole, *Shaping a City*, 183; BCCAR 1961–62, 22.
25 Report of Melbourne and Metropolitan Board of Works 1951 (RMMBW), 20.
26 RMMBW 1953, 18. 27 RMMBW 1953, 18–19. 28 RMMBW 1951, 22.
29 RMMBW 1955, 14–15.

At the start of the 1950s, 95 per cent Melbourne's sewage and trade waste was purified and disposed of at the Werribee farm. This waste water was purified through land filtration, with sales of the cattle that grazed on the farm helping to offset costs, with the treated effluent drained into Port Phillip Bay.[30] With the volume of sewage to be treated exceeding the farm's capacity, the Board agreed in 1953 to discharge untreated sewage into the ocean at Cape Schanck.[31] Board investigations in 1956 found that the Werribee farm would reach capacity by about 1970.[32] The Braeside treatment works was opened in 1940; by 1956, it was processing twice the amount of waste for which it had been designed. At this time, Melbourne's rapid industrial growth meant that 15 per cent of the sewage flow was trade waste, estimated to be 'equivalent of an additional population of 400,000 and on the basis of organic load on a sewage treatment plant, 750,000'.[33] Writing on the Braeside treatment works in 1956, the Board stated: 'the limit of effective operation has now been reached', but the same could have been said for the entire system.[34] Between 1945 and June 1957, the Melbourne Board supplied water to an additional 131,428 additional dwellings, sewerage to 74,971, and constructed 74 km of drains. However, in September 1957 the Board reported that the 'existing water supply and sewerage systems are so overloaded that a major breakdown in either of them could have serious consequences'.[35]

In April 1960, the MMBW reported a backlog of 85,000 unsewered holdings. The system now served a population of 1.375 million, causing mains to be overloaded.[36] The Board was fighting a losing battle, as the population it served was projected to increase by 25 per cent, to reach 1.75 million by 1970.[37] In 1959–60, 549 septic tanks were emptied at houses, and over four million nightsoil pans were delivered to depots.[38] A new south-eastern interceptor and disposal works, approved in 1955,

[30] *Metropolis Quarterly Journal of the Melbourne and Metropolitan Board of Works (Metropolis Quarterly)*, 1956, 8; RMMBW 1951, 23–24. Dry years reduced the stormwater load. In 1961–62, the daily average load was 5.5 per cent less than that of the previous year. In summer, when evaporation was high, most of the water at the farm was used to water pasture grasses, with less flowing into Port Phillip Bay. RMMBW 1956, 1–2, 13.

[31] RMMBW 1953, 19.

[32] *Metropolis Quarterly*, June 1956, 4.

[33] *Metropolis Quarterly*, June 1957, 8.

[34] RMMBW 1955, 14–15.

[35] *Metropolis Quarterly*, September 1957, 4.

[36] *Metropolis Quarterly*, April 1960, 13.

[37] *Metropolis Quarterly*, April 1960, 9.

[38] *Metropolis Quarterly*, October 1960, 17.

was to be completed by 1970; the south-eastern sewerage system was under construction by the early 1970s.[39]

In Sydney, the Cumberland County Council, a new authority sitting between local councils and the State government, completed a master plan in 1948 that included provision for sewerage, street paving and drainage, and other infrastructure in new suburbs. By the time the plan was implemented in 1951, the metropolitan population was growing at twice the projected rate and much unplanned suburban development had taken place.[40] In 1951, 30 per cent of houses in metropolitan Sydney were not connected to sewerage; ten years later, the proportion had increased to 33 per cent.[41] In 1957, a journalist reported that in Sutherland Shire, where the population increased nearly five-fold from 29,000 in 1947 to 132,000 in 1966, the lack of sanitation, road paving, and drainage would 'shame an outback town'.[42] The investment required for the MWSDB to do so, estimated at $2,000 million in 1969, far exceeded the Board's annual capital expenditure of $40 million. In 1968, economist Edwin Walder maintained that ad hoc construction without systematic long-term land use planning had wasted public capital investment and created a seemingly insurmountable backlog of needed sewerage works.[43]

Elected Brisbane City Council Lord Mayor in 1961, Clem Jones took over a city where night carts still operated in 80 per cent of the suburban area. A surveyor by training and a prodigious developer of suburban land for sale, Jones convinced the State government to amend the *City of Brisbane Act* to allow the creation of Sewerage Project Areas and compel the Queensland Housing Commission and private developers to contribute to a sewerage trust fund.[44] In 1961–62, his first year in office, a record 3689 houses were connected to sewerage.[45] Between the end of 1963 and March 1967, the proportion of Brisbane that was sewered increased from 40 to nearly 60 per cent.[46] In 1963, 8203 truckloads of night soil were delivered (32 per day); by 1967, this had decreased to 3729 loads (10 per day).[47] Construction peaked in 1969–70, when 15,033 new sewerage connections were completed; the system now served an estimated half-million people. Septic tanks continued to be built, but these were limited to new estates beyond the reach of sewerage and only amounted to

[39] *Metropolis Quarterly*, December 1956, 6–7.
[40] Freestone, *Urban Nation*, 145–48.
[41] Spearritt, *Sydney's Century*, 94.
[42] Cited by Ashton, Cornwall, and Salt, *Sutherland Shire*, 157.
[43] Walder, 'Water and sewerage'.
[44] Cole, *Shaping a City*, 196.
[45] Cole, *Shaping a City*, 197.
[46] Cole, *Shaping a City*, 200.
[47] BCCAR 1962–63, 28; BCCAR 1966–67, 29.

32,650, 16 per cent of the metropolitan housing stock. Only 9139 properties in sewered areas were not connected.[48] New sewerage lines, pumping stations, and sewerage treatment works were designed by the BCC throughout the city, including private estates, with work completed by contract and funded by BCC, State government subsidies, and developers, including the government entities of the Queensland Housing Commission and War Service Homes.

Work began in 1973 on a water treatment plant at Luggage Point, marking the end of disgorging untreated water into Moreton Bay, as well as the Eagle Farm pumping station augmentation program.[49] By 1973–74, when the BCC was spending $9.3 million per year to make the remaining connections, fewer than 5000 properties required nightsoil collection services. When newly elected Labor Prime Minister Gough Whitlam introduced a National Sewerage Program in 1973, with a 60 per cent subsidy on sewerage reticulation, an estimated 71 per cent of Brisbane's metropolitan population lived in sewered premises.[50] Ironically, having already largely sewered Brisbane, Clem Jones found his city ineligible for Federal funds. Whitlam offered to subsidise the treatment plant, but the incoming conservative Fraser government abolished the concession.[51]

As the five cities grew, with people renting or building detached houses, many of the new subdivisions incorporated what had previously been discrete country towns, where household water tanks were common. With the coming of town water, established houses were gradually connected to the reticulated system. The notion that anyone would want to retain tanks, often rusted and seen as a breeding ground for mosquitos, let alone install a new tank, faded completely from public view. Plumbers frowned on them, and municipal councils would not countenance them in new developments. Town water was the expected standard, along with coal-powered electricity; tanks were seen as a remnant of an earlier, primitive era. With the convenience of indoor bathrooms, flush toilets, and constant hot water, average daily consumption in the three east coast capital cities in the 1950s and 1960s was most commonly in the range of

[48] BCCAR 1969–70, 32; Commonwealth of Australia, Census data.

[49] Cole, *Shaping a City*, 203.

[50] BCCAR 1972–73, 37. The Whitlam government spent $330 million on the National Sewerage Program before it was cancelled by the Fraser government. During the life of the program, the backlog of unsewered properties in Sydney was reduced by 40 percentage points and in Melbourne by 45 percentage points.

[51] Cole, *Shaping a City*, 205–06.

300 to 400 litres per day, only occasionally tipping over 500 litres in Brisbane and Sydney.[52] Unlike their parents' generation, Australians born after World War II grew up assuming that water and electricity were services that they had to simply turn or switch on. While households paid for electricity on the basis of usage, water charges were usually a flat fee per property, without any extra payment for excess use. The notion of excess usage had little currency with seemingly endless suburban water supplies.

STORING WATER: DAMS AND RESERVOIRS

Following international precedent, especially from the United States, governments decided that the best way to secure a potable water supply for the growing metropolitan areas was to build dams and reservoirs – the bigger the better. Dependence of dams has a long history, linked with human desires to conquer nature and reap the benefits. The rise of civil engineering, firmly entwined with notions of progress, order, and civilisation, offered solutions to a range of social and environmental challenges.[53] Political scientist and anthropologist James C. Scott argues that this secular faith in engineering to control nature and provide social order reached new heights among modern states in the twentieth century.[54] Australian dam construction was closely aligned to notions of nation building, progress, and ensuring the health of the suburbs.

In 1941, Sydney's MWSDB engaged Wisconsin-trained civil engineer, Dr John L. Savage, to assess the Warragamba Dam site, catchment area, and the proposed mode of construction. Employed by the US Bureau of Reclamation, Savage had supervised the design of the Hoover and the Grand Coulee dams and advised the Chinese government about damming the Yangtze River. He approved the choice of the site and explained how ice and chilled water could be used to reduce concrete cracking, a major issue for large dam walls.[55]

The Sydney press could not resist the temptation to comment that 'Sydney will be assured of a far better water supply than any other Australian capital', pointing out that Melbourne's heatwaves 'gravely depleted' the city's limited reserves, while Adelaide still awaited an 80 km pipeline from the Murray River. Of the vast 9000 square kilometre

[52] Calculated from data in MWSDB, MMBW, and BCC Annual Reports.
[53] See for example Mitchell, *Rule of Experts*; Sofoulis, 'Big water'; Blackbourn, *Conquest*.
[54] Scott, *Seeing Like a State*. [55] Aird, *Water Supply*, 109–10.

catchment, 30 per cent of the lands nearest the dam were designated 'restricted special areas'. Much of this area came in the form of bushland, including the Blue Mountains National Park. Beyond this pristine area, the catchment included large tracts of cleared farming land and small and large towns, from Goulburn and Katoomba to Lithgow.[56]

When the Board planned the dam, it predicted that Sydney's population would rise to four million people by 2000, which turned out to be only a slight overestimate.[57] Such was the Board's concern about Sydney running out of water in the early 1950s – after the protracted dry conditions in the 1940s, hydrologists and engineers feared another drought — that the Board diverted most of its available funds into the dam, putting extensions to sewerage on hold.[58] New South Wales school children were told that 'the degree of civilisation of a country can be assessed from the annual consumption of water per head of population, not least because of the amount of water required to sustain manufacturing, from steelworks and dye works to food industries'.[59]

In 1959, a year before the Warragamba Dam was completed, there were a handful of articles about how the 'austere beauty' of the sunken valley – its floor below sea level – would 'be seen, no longer'. Journalist Jack Kenny penned a reflective piece on the history of the valley, 'a favourite camping and hunting ground for aborigines', citing the memoirs of William Russell, chief man of the Gun-dun-gora people, who died in 1914, aged 84. Russell recalled that as a young man he had camped with his people on the '"Big Flat", Burru-gu-rang, with kangaroos in great numbers'.[60] Indigenous sites were flooded by the dam, but in the absence of legislation to preserve them, and awareness of the science of archaeology some decades away, this occasioned almost no comment. The site of Aboriginal Reserve No.26, built on a large Gundungurra campsite, remains above the current high-level mark of Lake Burragorang. Stone tools were uncovered and identified there in 1988, in the bicentenary year of settler Australia.[61]

The Board rested on its laurels for some decades after the opening of Warragamba. Apart from the 1965–67 drought, the big dam lived up to its reputation to provide water for the rest of the century. Consumption

[56] Aird, *Water Supply*, 106. A detailed map of the catchment is included as endpapers to this 347-page, authoritative commissioned history.

[57] At the 2001 census, the population of the Sydney Statistical Division was 3.948 million.

[58] *Sydney Morning Herald*, 14 February 1952.

[59] *Water: Key to Development.*

[60] *Sydney Morning Herald*, 3 January 1959.

[61] Smith, *Aboriginal People*, 324.

rose to over 500 litres per capita by the late 1970s.[62] Reticulated water continued to be supplied to the rapidly growing suburbs to the north, west, and south of the city, and funds from the Whitlam government saw a sharp increase in the number of houses connected to sewerage. With water from toilets, laundries, and kitchen sinks all going into the sewers as septic tanks became redundant, the ocean outfall sewers saw a huge increase in volumes. Developers were forced, in new planned estates, to guarantee both water and sewerage connection.[63]

Like Sydney, Melbourne strove hard to work out how to store enough water to cope with rapid suburban expansion and summer heat waves. In 1941, the MMBW followed the example of the MWSDB, consulting Dr Savage, who recommended increasing the wall height of Eildon Weir across the Goulburn River to increase its capacity. By the late 1960s, the Melbourne supply system had a combined yield of 341 million cubic metres, sufficient for a population of 2.5 million, with an approximate daily per capita usage of 341 litres.[64] Upper Yarra Reservoir, opened in November 1957, could hold 97 million cubic metres, more than twice the combined storage of the Yan Yean, Toowourrong, Maroondah, O'Shannassy, and Silvan reservoirs.[65]

On completion in 1959, Brisbane's Somerset Dam had an approximate capacity of 200 million cubic metres at full supply level (twice the volume of Sydney Harbour) and was expected to supply 90 per cent of Brisbane's water until 1980.[66] Within a decade, Brisbane's suburban growth again stretched water supply capabilities. Preliminary planning work for Wivenhoe Dam, on the Brisbane River, 80 km from the city centre, commenced in the 1960s, and land acquisitions began in the early 1970s.[67] When floodwaters reached a height of 5.45 m at the Brisbane GPO in January 1974, the biggest flood in eighty years, the Wivenhoe project, planned not only to provide water but also flood mitigation, was expedited.[68] While the nearest third of Sydney's Warragamba's catchment is primarily bushland, almost all of Brisbane's Wivenhoe and Somerset catchments are agricultural, held by thousands of privately

62 Crabb, 'Coping with change', 212–13.
63 See Neutze, *Funding Urban Services*, 82–88.
64 *Metropolis Quarterly*, December 1957, 3.
65 *Metropolis Quarterly*, April 1959, 10.
66 Queensland Government, *Somerset Dam*, 12.
67 Grigg, *Comprehensive Evaluation*, 52.
68 For detailed accounts of Brisbane Dams, see Cook, *A River with a City Problem*; Cook and Spearritt, 'Water forever'.

owned farms, with the city's water supply subject to contamination from pesticides, fertilisers, and animal excrement.

TURN GRASS INTO LAWN

In the 1950s and 1960s, most suburban building sites were cleared of vegetation, so new home purchasers had to work out how to landscape their gardens. The vast majority planted lawns, still the evergreen choice for Australian households. These became much easier to 'tame' from 1952, with the invention and popularity of the Victa motor mower. Invented in the Sydney suburban backyard of Mervyn Victor Richardson, the mower did not use the scissors action of hand mowers but tore into grass, bracken, or anything else in the ground, including weeds, with a high impact rotary action. While the cut was coarser than earlier mowers, Victa lived up to its advertising claim to 'turn grass into lawn'. Households were encouraged to trade in their old hand mowers for the new petrol-powered machine.[69]

In the late 1950s, it was hard to pick up an Australian newspaper or a magazine that did not have an advertisement for Victa Mowers. In 1959, the 'Standard' at 49 guineas (£51/9) and the 'Special' at 55 guineas (£57/15) were beyond the reach of many households, when the Commonwealth basic wage was £12/6 per week, but newspaper advertisements gave details of trade-ins and hire purchase.[70] In popular magazines, the advertisements, increasingly in colour, offered instant lawns of which one could be proud. A full-page advertisement in the *Australian Women's Weekly* in November 1956 shows a housewife deploying the household Victa, saying 'we keep our lawns velvet smooth', while in another image a husband and wife proudly tell readers 'we cleared our land of Jungle Growth' (Figure 6.5). The rhetoric reflected the persistence of 'pioneering' concepts of civilising the savagery of the suburban frontier, one lawn at a time.[71]

A Victa mower could keep a lawn neat, but only regular watering could keep it green. The widespread use of sprinklers made that easy, especially when in most cities the idea of charging for excess water usage had not yet occurred to water authorities or was politically unpalatable. Indeed, in some suburbs water usage was not even measured at all, as the

[69] Hunt, 'Lawn mower', 364. [70] 'Tax Commissioner's report', *Tribune*, 27 May 1956.
[71] Gaynor, 'Lawnscaping Perth', 4; Hamer, *New Towns in the New World*, 206–19.

FIGURE 6.5. Keeping lawns green and 'velvet smooth' in 1950s suburban Australia (Are Media Pty Limited, aremediasyndication. *The Australian Women's Weekly*, 1956).

lag in installing water meters could take years to overcome.[72] The use of potable water for lawns was actively encouraged by magazines such as

[72] Spearritt, 'Water crisis', 20–21.

Home Beautiful. Journalist Olive Mellor wrote in a 1967 article, 'you can't beat a real grass lawn', as nothing is 'as restful to the eye as grass – or as comfortable to walk on', although conceding this 'green carpet' required 'considerable upkeep'.[73]

A typical arrangement of a lot in the 1950s and 1960s, with a just-built house on the urban fringe, included flower plantings and lawn in the front yard, with the house often placed about one-quarter of the way down. In the backyard a larger lawn, and another Australian invention, the Hills Hoist rotary clothesline (developed and manufactured by Lance Hill in his Adelaide backyard in 1945) was accompanied by the backyard toilet. In established suburbs closer to city centres, sewerage was more likely to be connected, and new homes were also more likely to have an indoor toilet. Many owner builders, with the growing acceptance and affordability of septic systems, decommissioned the backyard dunny and moved it indoors, usually separate from the bathroom, with a shower head above the bath. Very few houses in the 1950s and 1960s had more than one toilet.[74] Bathroom leaks were common, because shower walls were rarely tiled, with fibro sheeting or painted masonite a much cheaper option.

Outside the home, the Prest survey had found that 41 per cent of Melbourne households in 1941 were engaged in some form of home food production. On larger suburban blocks, 88 per cent of households produced food, a number that diminished in inner suburbs. The survey found that production was more common on owner-occupied land in middle-class or skilled working-class households.[75] Home growers cited 'thrift, leisure and food quality' as motivating factors, with contemporary garden magazines emphasising the economic benefits. Wartime shortages and a 'Grow Your Own' campaign linked home gardening with patriotic duty to increase food supplies. Post-war gardening magazines still advocated vegetable beds and fruit trees in the modern garden, albeit on a smaller scale and largely relegated to the backyard.[76] Many migrants, especially from Greece and Italy, were practiced in growing their own food and continued that tradition in Australia. Growing their own vegetables gave

[73] Olive Mellor, 'You can't beat a real grass lawn', *Australian Home Beautiful*, September 1967.

[74] The first A. V. Jennings home to offer an ensuite bathroom, off the main bedroom, in addition to the usual bathroom arrangement of bath/shower/vanity in one room, with separate lavatory in a smaller adjoining room, came onto the market in 1968. See Garden, 'The changing space', 140–43.

[75] Gaynor, *Harvest*, 101. [76] Gaynor, *Harvest*, 104, 132.

them tomatoes and lettuce but also other varieties almost unknown to Anglo-Australians, including basil and zucchinis.[77]

From the 1960s, planting gardens with Australian native species and doing away with lawns became more common in some new suburbs, thanks to a growing concern for wildflower conservation and the popular designs of Edna Walling. New middle-class housing developments on Sydney's north shore, especially abutting bushland, were often on steep sites that did not lend themselves to lawns. Native gardens were also developed in Melbourne's new bayside suburbs, where tea tree fences were common. The majority of neighbours still preferred the traditional lawn, but native plantings – often regarded as 'untidy' and reflecting a lack of interest in lawn mowing on behalf of the senior resident male – were tolerated. Although at the time few households had any thought that native gardens required less water, the likes of Walling encouraged a conservative approach to watering:

> Thousands and thousands of gallons of water are wasted each year on unnecessary watering. A well-mulched garden will stand up to a great deal of heat before it requires watering, if at all, and watering should *never* be started before it is absolutely necessary, for once you start, the resistance of the plants is weakened, and they no longer survive without this artificial aid.[78]

Nevertheless, with suburban residents using cheap water for lawns and gardens, car washing, and swimming pools, Melbourne authorities estimated that outside water use had increased by 52 per cent in the 1970s.[79]

WATER RESTRICTIONS

In July 1938, when Sydney was in severe drought, the MWSDB had responded to low levels of water supply by prohibiting fixed hoses and mechanical sprinklers for garden watering and appealed to households to be frugal in water use. This was a time-worn approach to water management. Australian governments and water authorities have traditionally used sprinkler regulations to curb residential water usage, a strategy justified by garden usage accounting for over half of household water consumption at least until the 1980s.[80] In the following month, the Board banned the washing of motor vehicles and postponed water and sewerage

[77] Seddon, 'Australian back yard', 23. [78] Walling, *A Gardener's Log*, 117.
[79] Dingle and Rasmussen, *Vital Connections*, 368. [80] Gaynor, 'Lawnscaping Perth', 2.

connections.[81] Households without water meters could pay a 'garden fee' for water usage. When illegal hose watering was exposed (without fee or meter), a comprehensive press and radio campaign was conducted and fines introduced; its success led to a decision in February 1940 to restrict the garden fee to hand-held hoses.[82] Water meters were compulsory for properties exceeding 1000 square feet (92.9 square metres), allowing unlimited water usage but payment for excess consumption, while smaller property owners could pay eight shillings annually for a garden fee to use a garden hose.[83] Falling dam levels forced further restrictions in 1940: first, the banning of fixed hoses and sprinklers in some parts of Sydney, then the prohibition of private hoses and swimming baths, showers at surf sheds, and street watering in August as the Nepean storage level fell to 28 per cent. Market gardeners and nurseries had to reduce their usage. Industrial users began to re-use water, saving 9 million litres per day.[84] When rain fell in December, restrictions eased, resulting in 'a more liberal use of water than had been anticipated'; restrictions were re-introduced in May 1941. Garden watering was reduced from four hours daily to one hour and 950 infringement notices were issued as the Board moved away from garden fees and towards meters and a user-pays approach for excess water.[85]

Confronting public criticism in 1941, the MWSDB justified the restrictions. The Depression, it claimed, had halted work on Nepean Dam for two-and-a-half years, causing the loss of '18,000 million gallons [82 million cubic metres] that would otherwise have been caught before the drought commenced'. The Board boasted that they had not run out of water, despite the admission that in a 'young country such as Australia where rainfall statistics do not yet extend over a century, knowledge of what really constitutes an all-time record is quite unsubstantial'.[86]

The dry conditions continued as Sydney's dams dropped close to 91 million cubic metres, which the Board regarded as a threshold for dangerously low levels. Watering gardens and sporting fields was banned, and primary industry could only use water for two-and-a-half hours each day. Household tap use was limited to four hours in the morning and afternoon and only sanitary flushing was permitted with bricks or bottles

[81] Metropolitan Water Sewerage and Drainage Board Annual Report (MWSDBAR) 1939, 13.
[82] MWSDBAR 1941, 32.
[83] 'When hoses can be used', *Sun*, 17 February 1940; MWSDBAR 1941, 29.
[84] MWSDBAR 1941, 33. [85] MWSDBAR 1941, 33. [86] MWSDBAR 1941, 37.

placed in toilet cisterns to use less water. Industries were asked to install circulating plants to re-use processed water. As total water reserves reached its 'nadir' of 70 million cubic metres (12 per cent of capacity), a 'phenomenal storm' added 59 million cubic metres to the storages, and restrictions on taps, toilets, and hot water were lifted immediately.[87] The restrictions were effective with Sydney consumption between October 1941 and March 1942 reaching a daily average of 198 litres per capita.[88] Heavy rains in October 1942 finally broke the drought that had prevailed since October 1934, and storage reached 495 million cubic metres – the highest in the Board's history. Recording its 'thankfulness to God', the Board declared that this placed 'the water supply position beyond all doubt', but within two years water restrictions were again in place.[89] In December 1944, fixed hoses and sprinklers could only be used in gardens with a meter. People without a meter could use a hose if they paid a 'garden fee' to use hoses; otherwise, only watering cans could be used.[90]

Adelaide's garden water consumption figures amounted to about half the city's potable water in 1949, and as water supplies dwindled, the State government introduced drastic restrictions. With sprinklers banned, hosing was limited to two periods of two hours each week. Sporting bodies could use 40 per cent of their normal summer consumption, and commercial and public gardens were issued with permits. These measures were estimated to reduce usage by a quarter. Works Minister Malcolm McIntosh boasted that Adelaide stored 159 million cubic metres per capita, compared with Melbourne's 86 million cubic metres, but this reflected differences in each city's climate. A drought for much of the early 1940s meant Adelaide's reservoirs had caught the least amount of water since the 1914 drought. When police reported those flouting sprinkler restrictions of four hours every evening in January 1950, the concern was not about saving water but rather the power load on the pumps extracting water from the city's artesian wells.[91] Despite the growing number of bores in use, without regulations Adelaide faced empty reservoirs by April 1950. Consumption increased and appeals for restraint went unheeded as consumers, according to McIntosh, 'would not make a voluntary contribution for the common good'. Hoses were banned and £20 fines introduced.[92] Adelaide residents continued to be

87 MWSDBAR 1942, 29. 88 MWSDBAR 1942, 30. 89 MWSDBAR 1943, 9.

90 'Home Garden Water Warning', *Sun*, 9 December 1944.

91 'Water restrictions well observed', *West Australian*, 5 January 1950.

92 'Severe water restrictions', *Advertiser* (Adelaide), 22 March 1951.

enthusiastic water users. In 1958, average allotments in Springfield, at 2136 square metres, were using 3288 litres per day, while the smallest allotments, in Hindmarsh and Bowden, at 522 square metres, were using 536 litres per day. The Town Planning Committee in 1962 hoped that domestic consumption would stabilise at 636 litres per head per day.[93]

By the mid-1970s, water consumption in Perth had tripled since the end of World War II, to the highest per capita level of any of the five cities.[94] Both public and private lawns were forever green. To meet demand, additional storages, an additional bore in the southern suburbs, and a dam on the Serpentine River near Jarrahdale were constructed. While the demand can be partially explained by the increase in water-consuming white goods and changing notions of cleanliness, the biggest culprit was an obsession with maintaining green lawns on sandy soils in Perth's dry summers for about five months per year.[95] Over half of Perth's water consumption was used on gardens, the proportion increasing to about 70 per cent in summer.[96] The nylon garden hose (introduced in 1945 by Nylex, a Melbourne manufacturer) and improved sprinklers encouraged profligate usage, prompting academic George Seddon to warn Western Australians to 'fear the hose', as 'once you start using the hose, your garden becomes dependent on it and you are hooked forever'.[97]

As Brisbane grew, water reticulation struggled to meet demand. Dams reached their lowest recorded levels in the history of Greater Brisbane in 1941, and hoses were banned between 6 am and 7 pm, 'saving a million gallons [4.5 million litres] daily'.[98] As the Brisbane River stopped flowing at Mt Crosby in January 1942, weir levels fell and the 'water supply position was gradually becoming more serious', the BCC placed a complete ban on water usage for watering (garden, lawn), including buckets. The *Telegraph* reported that the restriction had an immediate impact, as water usage fell and the Mt Crosby Weir water level rose by 5 cm.[99] Sprinkler restrictions were lifted in April 1949 but again introduced in

[93] *Report on the Metropolitan Area of Adelaide* (RMAA), 1962.
[94] Morgan, *Running Out?* 82; Gregory, *City of Light*, 38.
[95] Morgan, *Running Out?* 83–84.
[96] Morgan, *Running Out?* 85; Williamson, 'Statistics', 20.
[97] Davison, 'Down the gurgler', 56; Morgan, *Running Out?* 86; Seddon, *Swan River Landscapes*, 49–50.
[98] 'Water reserves fall', *Sunday Mail* (Brisbane), 14 December 1941.
[99] 'Total Ban on Water for Garden', *Telegraph* (Brisbane), 30 January 1942.

1951–52 during a period of drought, reflecting a vacillating pattern of restrictions and abundance.[100]

One weekend in August 1951, the BCC received widespread complaints from suburban residents had lost water pressure. Dry weather, heavy garden watering, and water use to fight fires created extraordinary demand, forcing the council 'to use everything we had to cope with it'.[101] Boosters on the Mount Crosby trunk mains (installed in 1950) had to be deployed to increase pressure. Despite the filtration plant and trunks mains working at 'full capacity', with daily consumption having increased from 123 to 159 litres per day in three years, the *Courier-Mail* reported that council was 'straining every resource to avoid hosing cuts this summer'.[102] After the driest winter in five years, consumption reached a record daily record of 178,661 cubic metres on 25 August 1951. As the suburbs of Wynnum, Manly, and Lota were left without water at times during the weekend, Council introduced water restrictions, banning sprinklers and enforcing hand hosing, except for market gardens and nurseries.[103] Consumption continued to increase, forcing an immediate sprinkler ban. Limited supplies of steel hampered the Council's efforts to meet water demand by expanding the city's reticulated network.[104]

With sprinklers banned in Brisbane since February 1952, on 9 August 1953 BCC lifted all restrictions for one Sunday, as a trial for households and dusty sports and school grounds. Council was confident that people, 'conscious of their responsibilities', would not abuse the privilege.[105] But they did, and at 7.30 am on the day that became known as 'Black Sunday', at least fifteen suburbs had no water, forcing Mayor Frank Roberts to make an urgent radio appeal at 9.30 am asking householders not to use sprinklers that day.[106] The *Courier-Mail* reported reservoirs being reduced to dangerous levels.[107] An all-time high was reached on 27 November 1958, when 259,127 cubic metres were used in one day.[108]

100 'Ban on garden sprays off now', *Telegraph*, 27 April 1949; Cole, *Shaping a city*, 138; 'Big drop in water use', *Courier-Mail* (Brisbane), 21 December 1951.
101 'Drain in city water', *Courier-Mail*, 13 August 1951.
102 'Council worried over water', *Courier-Mail*, 4 January 1951; 'Drain in city water', *Courier-Mail*, 13 August 1951.
103 'Ban fixed hoses, sprinklers', *Courier-Mail*, 28 August 1951.
104 'Sprinkler ban on after record day', *Courier-Mail*, 24 September 1951.
105 'Householders, schools and sporting clubs benefit', *Courier-Mail*, 4 August 1953.
106 'Chaos follows removal of sprinkler ban', *Townsville Daily Bulletin*, 10 August 1953.
107 '150 "caught" breaking sprinkler, hosing rules', *Courier-Mail*, 15 December 1954.
108 BCCAR, 1958–59, 35.

Melbourne experienced heat waves in the summer of 1950–51 that stretched the capacity of water supplies. From 3 pm until 9 pm on 6 December 1950, many homes were without water as the North Essendon Reservoir ran dry. Concerns arose over two additional reservoir supplies as the chair of the MMBW expressed concern that the drain was so great that the 'whole supply could be endangered'.[109] All watering, including parks, was banned. The heat waves demonstrated Melbourne's extreme water consumption patterns. On a hot day, Melbourne consumed 4.4 times more than on a cold day, more than twice the average daily consumption for the year. A worldwide survey in 1949 gave Melbourne the dubious title of the highest ratio of maximum to average usage of any major city. On a hot day, Melbourne used 2.13 times the average daily consumption, compared with Sydney (1.61), London (1.24), New York (1.25), Paris (1.21), and Chicago (1.32).[110] Australia might have been the driest inhabited continent, but its capital city residents expected to water their lawns whenever they wanted to, and with the provision of abundant cheap water supplies, their governments obliged.

Melbourne's problem of excessive summer usage was compounded by the increasing number of new properties requiring piped water. Consumers were asked to 'practice economy' and eliminate garden watering in later afternoons and evenings. As temperatures rose in December 1950, new by-laws allowed the Board to make anything beyond hand watering illegal for five days. When reservoirs continued to fall in 1951, lawn watering was banned. More draconian measures saw temporary water restrictions introduced in 1952.[111] Despite a sprinkler ban in August 1953, in many places flow was reduced to a trickle. The problem had been exacerbated by City Council parks and private gardens 'using water almost non-stop in an effort to have Melbourne looking its best for the Queen's visit' in a hot summer.[112] The Royal Visit in 1954, the first by a reigning monarch, saw Australians – in cities and towns – putting themselves on display as never before.

A system of booster pumps was installed in Melbourne until steel could be supplied to build additional mains, as the additional 146 km laid in the previous twelve months (1950–51) did not meet demand.[113]

[109] 'Near-danger level in water supply', *Age* (Melbourne), 7 December 1950.
[110] *Metropolis Quarterly*, April 1959, 3. [111] RMMBW 1952, 11.
[112] 'Melbourne has sprinkler ban', *Morning Bulletin* (Rockhampton), 9 January 1954.
[113] RMMBW 1951, 19–20.

Consumption had increased by 40 per cent since 1939, as almost 77,000 additional properties had been supplied with reticulated water. Despite building 1280 km of pipes, the Board could not keep up.[114] The difficulty in securing loans led the Board to declare that, at the existing rate of mains construction, 'it will be some years before the back-lag of reticulation will be overtaken'.[115] By June 1957, 440,159 properties had been supplied with water, serving a population of almost 1.6 million.[116]

The solution followed the same well-worn path, construction of the Upper Yarra Dam, approved in 1946. Opened in late 1957, the Reservoir would provide an additional 500,000 kilolitres daily to the metropolitan area, making daily supply 1.35 million cubic metres for a city that consumed just over a million cubic metres per day.[117] But dams would not solve distribution problems, as the Board reported low pressure and water shortages in times of peak demand at the end of the almost 1300-square kilometre distribution system.

Melburnians' love affair with their gardens continued. Watering lawns and gardens, according to the MMBW in 1960–61, was the 'biggest strain on the water supply system', as it bemoaned residents' practice of a shallow sprinkle every day rather than a weekly soak. The Board considered that a single household of two (without a garden) should use about 25,000 gallons (113,652 litres) annually, with 40,000 gallons (118,844 litres) for the average lawn and garden an acceptable level.[118] Melbourne experienced the 'Year of the Great Drought and Water Crisis' in 1967–68, according to the Board, 'adding urgency to the need for augmentation of the water supply system'.[119] Emergency works were constructed to divert additional water from the Yarra, Thomson, and Tanijl rivers into the metropolitan system, with restrictions introduced.[120] In response to the drought, construction began on the 273 million cubic metre Cardinia Reservoir in 1969–70.[121]

Drought conditions in the first and last quarters of 1970 in Brisbane caused record annual water consumption of 142 million cubic metres in a single year, despite prolonged rainfall from November to March 1971.[122] Somerset Dam withheld a flood, the spillway gates used for the first time since the 1955 flood. Work commenced in 1970 on a 200,000 cubic metre

[114] RMMBW 1952, 19. This work relied on large loans: £12 million (1916), £28 million (1946), and £59 million (1956–57). *Metropolis Quarterly*, June 1957, 13.

[115] RMMBW 1953, 16.

[116] *Metropolis Quarterly*, December 1957, 21.

[117] RMMBW 1951, 11–12, and 1952, 12.

[118] *Metropolis Quarterly*, October 1960, 9.

[119] RMMBW 1967–68, no pagination.

[120] RMMBW 1967–68, no pagination.

[121] RMMBW 1969–70, no pagination.

[122] BCCAR 1970–71, 18.

capacity North Pine River Dam, constructed by the Co-ordinator General's Department on behalf of the BCC.[123] The first major water storage project since Somerset Dam, it first stored water in May 1974. By June 1974, Brisbane's water system had thirty-two service reservoirs (with a capacity of 0.5 million cubic metres) and 4454 km of mains in use.[124] The new dam immediately encouraged residential development, as noted in the BCC annual report: 'As the dam nears completion, subdividers are becoming active in the catchment'.[125] As urban growth increased, water demand and the flood hazard followed.

A particularly hot Perth summer in 1971–72 led to the Water Board's implementation of a voluntary water conservation campaign between December 1972 and March 1973. Attempts to encourage more frugal use by publishing details of daily consumption in the *West Australian* newspaper had little effect. Restrictions were made mandatory in January 1974, voluntary in late 1976, and again mandatory from July 1977 to October 1978. Neither a public education campaign nor fines for flouting the restrictions could force a lasting decrease in consumption.[126] Cheap water did not help. In 1973, at 28 cents per thousand gallons, Perth's water was among the cheapest piped water in the world. By comparison, the price was 40 cents in Adelaide and Melbourne, 45 cents in Sydney, 44 cents in London, and 90 cents in West Germany.[127]

Water restrictions were electorally unpopular. Journalist Rohan Rivett warned the Victorian Liberal State government of backlash from the 'tens of thousands of young marrieds in the eastern and south-eastern suburbs, each with their own little cherished garden'. Only bath and shower water could be used on lawns and grass strips.[128] Industry continued to use 35 per cent of Melbourne's potable water, 'most wasted in cooling processes where water would be adequate'.[129] Water shortages never seemed to remain in the public memory for long. A year after the 1973 water rationing, journalist Lawrence Money reported:

> right now during summer heat the average Melbournian is splashing water on to his garden and lawn with gay abandon. His car, freshly washed and copiously

[123] BCCAR 1968–69, 21. [124] BCCAR 1972–73, 26.

[125] BCCAR 1972–73, 27. Work underway also included preservation of Samsonvale Bora Ring.

[126] Morgan, *Running Out?* 114, 120; Morgan, 'A thirsty city', 82.

[127] Athol Thomas, 'WA threatened by water shortage', *Canberra Times*, 1 January 1973.

[128] Rohan Rivett, 'Political danger: Backlash from water lack in Melbourne', *Canberra Times*, 3 February 1973.

[129] 'Water, water everywhere…', *Tribune*, 30 January 1973.

rinsed, sits in the drive. It's as though last year's near catastrophic shortage of the city's lifeblood never occurred.[130]

On 12 January 1973, after a long period of water restrictions, Melbourne had only eighty-four days of water left. The restrictions were lifted on 4 April, after 166 days, a period exceeded only during the 1967–68 drought, when they were in force for 354 days. The government, according to Money, was jolted into action, launching a $187 million dam-building program to build Yarra Brae Dam, Sugarloaf Creek storage, and Stage 3 of the Thomson River project.[131]

A BLUNT INSTRUMENT?

Historically, water restrictions have proven to be a 'cheap, quick and effective strategy' to reduce consumption, but they are also 'an especially blunt instrument because it does not differentiate between users and uses of household water'.[132] Restricting supply is a short-term, reactive policy to a crisis, rather than a considered, planned response to water usage. What is an appropriate use for potable water? What are appropriate consumption levels? Restrictions do not address the path dependency of reliance on large-scale water storages, engineering structures designed to increase supply. Nor do they create long-term cultural change to conserve a limited resource. Instead, this strategy supports a 'boom and bust' mentality of water usage, resentment towards water restrictions, and dependence on increasing supply.

Some citizens did appreciate that households could moderate their demand for water. In the 1940s, people moving from the country, where they had relied on water tanks, were often amazed at how city residents would take very long showers and water their gardens whenever they chose, rather than at optimum times for water retention.[133] While statutory authorities that controlled water supply and household billing always counselled residents and industry to reduce water usage at times of drought, the messaging waned promptly when heavy rain replenished dams and reservoirs.[134] Statutory authorities obtained most of their

[130] Lawrence Money, 'Supplying water to Melbourne', *Canberra Times*, 7 February 1974.
[131] Money, 'Supplying water to Melbourne'.
[132] Morgan, *Running Out?* 180; Troy, 'Life and times', 17.
[133] 'Every drop counts', *Telegraph*, 12 October 1940.
[134] See for example the media campaign in Sydney 1940 and 1941. 'Save water now', *Sun*, 21 August 1940; 'Easy to save water', *Sun*, 9 September 1940; 'You can save water', *Sun*, 10 September 1940.

income from selling water and charging for sewage disposal, the latter charge often based on the assumption that the amount of water coming into the property could be used as a surrogate for the amount of effluent leaving the property. It would be some decades before the idea of reducing water consumption year-round became an explicit tool of public policy. When it did, few water authorities could resist the temptation to encourage residential consumers to relax their vigilance if the dams were full.

7

Crises of Confidence

On the eve of the 1978 State election in New South Wales, incumbent Labor Premier Neville Wran declared, 'my intention is to run the state like a business. If it is sloppy and inefficient, then I want to get rid of the sloppiness and inefficiency.'[1] After the so-called Wranslide, Wran's government undertook to review the state's administration, and the MWSDB was among the first to come under scrutiny. US management consultants McKinsey and Company, which had provided advice to Australian governments on public relations and budgetary concerns since the early 1970s, advocated cuts to the Board's spending on both its staff and future projects.[2] Since over 90 per cent of the Board's area was now sewered and the recent completion of the Shoalhaven scheme ensured sufficient water storage for the city until the 1990s, the consultants reasoned, any further projects would be only 'qualitative' and 'not produce revenue'.[3]

In the wake of this report, the Secretary of the Water and Sewerage Employees' Union advocated a new direction for the city's water management:

> Just as the 1950s was the era of increased water supply, and the 1960s and 70s that of overcoming the sewerage backlog, so the 1980s ought to be that of attacking the pollution of our city. For the sake of a few cents extra from each resident our rivers and streams can be brought back to life and our beaches

1 Peter Kennedy, 'Making the cash go further', *Sydney Morning Herald*, 29 March 1979.

2 See Wilenski, *Directions for Change*.

3 Alan Mitchell and Ross Gittins, 'Water Board faces drop in staff', *Sydney Morning Herald*, 27 December 1978.

restored to their natural beauty. It will never be easier or cheaper to perform these tasks than now.[4]

Although the Secretary had his members' interests in mind, his assessment of the Board's past, present, and future pointed to the changing priorities of water managers in the five cities.

In the years after World War II, until the end of the 'golden age' of economic growth in the early 1970s, the provision of urban infrastructure by governments, although frequently lagging demand, had not been perceived as problematic.[5] This was about to change. The decades following post-war reconstruction and the push for increased immigration had confirmed governments' responsibility for all stages of the water network: from the construction of large dams and treatment plants, through infrastructure maintenance, water quality, and sanitary standards, to responding to emergency leaks in suburban streets and billing users. To achieve high quality standards across the entire system required large numbers of engineers, scientists, planners, tradespeople, and labourers, all coordinated by state government departments.

Their achievements are easily overlooked. In 1947, roughly half (3.8 million) of Australia's population lived in the five cities; by 1971, the figure was 7.5 million.[6] While population had almost doubled, the proportion of private urban dwellings supplied with reliable, clean water remained at over 98 per cent.[7] The construction of major water storages and networked systems in each city meant that even in one of the driest capitals, Adelaide, per capita water consumption had grown from 317 litres a day in 1947 to 487 litres a day in 1975.[8] Largely invisible to users due to their relative affordability and accessibility, these expanding water supplies were facilitating escalating consumption across the Australian suburbs.

While almost every city dweller had a supply of acceptable drinking water, the provision of adequate sewerage infrastructure continued to lag behind demand. The problems associated with State governments' inability to keep pace with population growth and new suburb expansion

[4] Bob Hunt, 'The disposal of sewage', *Sydney Morning Herald*, 4 January 1979.

[5] McLean, *Why Australia Prospered*, 176–209. [6] Frost, 'Urbanisation', 249.

[7] Calculated from Census of the Commonwealth of Australia, *Dwelling Summary Bulletins* for New South Wales, South Australia, Western Australia, Victoria, and Queensland, 1947, and Census of Population and Housing 1976 Census, *Occupied Private Dwellings by Source of Water Supply*, supplied by ABS.

[8] Calculated from South Australian Government, E&WS Annual Reports (various years) (Adelaide: Government Printer).

continued through the 1960s and 1970s. By the late 1960s, over 10 per cent of all Australians lived in dwellings without sewerage connections.[9] Authorities had been slow not only to connect many households but also to improve ageing treatment plants and outflows. In most of the five cities, sewage, after varying levels of treatment, was emptied into the sea through outlet pipes of sometimes inadequate length. In Sydney, it would not be until the late 1980s that the authorities, under public pressure, began to address the issue.

Although the urban water supply and waste removal authorities were properly criticised for being slow to meet the demands of new suburbs or to respond to the problem of waste disposal, they were also, in part, victims of their own success. Comparatively cheap water and sewerage services were 'taken for granted' by many residents, who – thanks to the convenience of large-scale, centralised networks of water provision and labour-saving devices – established lush green gardens, removed old water tanks, and continued to increase their annual consumption of water. Whether the price they paid for such services actually covered the cost of supply was not considered to be an issue; government services were there to ensure equitable and cheap supplies that would facilitate the fulfilment of suburban aspirations. Increasing services charges was resisted by both governments and citizens. While a lack of water to the household would cause an immediate outcry, the less obvious and more general problem of an inadequate waste treatment system initially raised fewer voices of protest. Being 'out of sight and out of mind' meant that the true costs of ignoring the environmental impact of existing systems were only slowly being revealed. It would not be until the issues became personal and visible, such as with raw sewage at popular beaches, that a public now attuned to ecological as well as aesthetic and health concerns began to complain and pressure the authorities to respond.

RATING WATER

Wielding new plastic garden hoses and with supplies assured by the post-war growth of water infrastructure, urban Australians embraced the

[9] ABS Census 1971 Bulletin No 7 (various issues). The proportions are for private dwellings, which includes houses (separate, attached, semi-detached, and terrace), villa, flats, and 'other'. Only 82 per cent of dwellings in Sydney were connected to the main sewerage system. In Melbourne, the proportion was 75 per cent, in Brisbane 72 per cent, and in Perth, which relied mostly on septic tanks, 43 per cent. Adelaide, which had long boasted of its lead in this field, had 88 per cent of dwellings connected to the main system.

suburban ritual of watering their gardens with gusto.[10] By the mid-1960s, the Australian Consumers' Association could observe, 'The amount of garden hose in Australia could, like Puck, put a girdle around the earth with very little difficulty.'[11] Ensuring suburban supplies spurred State governments to develop further water resources – as Perth engineer Ian O'Hara noted in 1967, 'It seems to be quite well-established that the garden suburb is part of the accepted way of life.'[12] In Melbourne, the record drought of 1966–67 expedited the development of the Cardinia Reservoir and later, in 1974, the Greenvale Reservoir.[13] In Perth, the North (1971) and South Dandalup (1974) rivers were dammed, becoming the largest of the reservoirs supplying the city.[14]

Despite such investments, the nation's water utilities were struggling to keep pace with the growth in demand. At this time, estimations of future demand tended to assume that population and per capita consumption would continue to increase at existing rates, and water would have to be supplied accordingly. The relatively high level of available water resources per head in Australia's cities reflected this approach.[15] As noted in Chapter 6, Perth householders became prodigious users of water after World War II, outstripping the residents of the other Australian capitals. Total water consumption in Perth had been doubling almost every fifteen years since 1920.[16] Calculations in the early 1970s indicated that this pattern would continue, such that demand would exceed available supplies by the mid-1980s.[17]

In a 1977 issue of the Australian Conservation Foundation's journal *Habitat*, George Seddon described Perth as an 'exceptionally vulnerable city' owing to 'its almost total dependence on stored water'.[18] Yet, the city was already shifting away from its reliance on dams. International efforts to advance global hydrological studies in the early 1960s had encouraged Australians to undertake the long-awaited measurement of the continent's water resources.[19] As part of its assessments of the underground water resources of Western Australia, the State's Public Works Department undertook exploration of Perth's groundwater reserves between Gingin

10 Davison, 'Down the gurgler', 56. 11 'Garden hoses', *Choice* 5 (1964), 140.

12 I. J. O'Hara, 'Utilization of water resources with particular reference to lawn and garden watering', Metropolitan Water Supply, Sewerage, and Drainage Board, Cons 6733, 1967/880672, State Records Office of Western Australia.

13 O'Hanlon and Spearritt, 'From water engineers', 98–112, 102. 14 Hunt, *Water*, 94.

15 Smith, *Water in Australia*, 191–92. 16 Hunt, *Water*, 96.

17 See Sadler and Field, *South West*. 18 Seddon, *Swan Song*, 54.

19 Boughton, 'Hydrology', no pagination.

in the north and Mandurah in the south. These studies revealed extensive stocks within two superficial formations about 70 m below the ground that the Water Board considered could be cheaply utilised for potable water supplies.[20] The MWB's development of these underground resources commenced with the commissioning of groundwater schemes on the Gnangara Mound with Mirrabooka in December 1970, followed by Gwelup (1974), Wanneroo (1976), and East Mirrabooka (1980) and on the Jandakot Mound in 1979.[21] By the mid-1970s, the aquifers provided about 10 per cent of Perth's water supply, and the city's reliance on this source was expected to grow.[22]

The development of groundwater supplies could not forestall the return of water restrictions in mid-1977. Implemented in the wet winter months for the first time, the tough restrictions lasted from July 1977 to October 1978 and were then eased to partial restrictions to May 1979. The MWB also embarked on a publicity campaign to reinforce the importance of water conservation and to show the severity of situation.[23] But water restrictions could not continue indefinitely, as they were proving both financially and politically costly to the conservative state government. The emerging influence of market principles in water resource management overseas led the state government to implement a 'user pays' water rating system in July 1978, which rewarded consumers who used less water.

Developments in Perth reflected a wider transformation that had unfolded over the course of the decade that other Australian water utilities would soon follow. In the United Kingdom and United States, water managers had become increasingly aware that water consumption was not merely for the purposes of public health – the 'sanitary revolution [was] over'.[24] A gulf had now emerged between needs and demands.[25] The cost of water provision had grown since World War II, as most of the cheapest sources had already been utilised and other public works were competing for these funds. The changing outlook on water provision was just one manifestation of a wider shift in public policy, whereby conservative governments around the world were reducing the size of their

[20] Cargeeg et al., *Perth Urban Water Balance Study*, 38.
[21] Pollett, 'Public water', 201, 226. [22] Allen, 'The Swan coastal plain', 14.
[23] 'Final water gauge', *West Australian*, 4 April 1977.
[24] Hancke and Davis, 'Demand management', 555, 559.
[25] Smith, 'Trends in water', 243.

bureaucracies in the face of expanding budgets, rising prices, and increasing unemployment.

Australian governments and water utilities recognised that they also faced challenges regarding water supply and demand and invited international experts to advise them on ameliorating these pressures. Perth's Metropolitan Water Board engaged the UK engineering firm Binnie International, as well as Eric Gilliland, financial director of the Thames Water Authority; a future adviser to US President Ronald Reagan, economist Steve Hanke of Johns Hopkins University; and Ken Townsing, former Under Treasurer of Western Australia. The resulting 'Binnie report' became the basis for the Metropolitan Water Board's plans to the end of the century.[26] Soon afterwards, the Hunter District Water Board in regional New South Wales engaged Binnie's expertise for their own assessment of their operations.[27]

These experts advocated a shift away from the development-oriented engineering approaches previously used in the management of water supplies. For Australian policy analyst John Pigram, managing water resources now required the creation of 'conditions which will bring about optimum use of water resources' through the 'adoption of measures to modify water needs and to maximise efficiency in water use'.[28] The primary tool to achieve this efficiency, argued the experts, was the price mechanism, that is, user pays. It was a policy that reflected the growing influence of neoliberal *laissez faire* thinking in Australian politics, which advocated the market over state intervention. In the wake of the political instability of the Whitlam government, the emergence of stagflation, and the well-publicised visit of Milton Friedman, conservative Australian politicians in the mid-1970s saw in these principles a viable alternative to what they perceived to be the failed Keynesian policies of the post-war era.[29]

Recognising that water was a valuable resource was not a new concept. Legal scholars such as Frank J. Trelease (1965) in the USA and Warren Musgrave (1974) in Australia had already argued that water as a scarce resource should be allocated using a price signal rather than government regulation.[30] This outlook received support in the international arena at

26 Binnie International, *Development Study*, 16–17.
27 See Lloyd, Troy, and Shreiner, *For the Public Health*.
28 Pigram, *Issues in the Management*, 300–01.
29 Moore, 'Life is not meant to be easy', 250, 255.
30 Trelease, 'Policies for water law', 1–48; Musgrave, 'Political economy'.

the 1977 United Nations Conference on Water in Mar del Plata, Argentina. This meeting, at which Australian representatives took part, produced the Mar del Plata Action Plan on the World's Water Resources, which emphasised among its recommendations that 'pricing and other economic incentives should be used to promote the efficient and equitable use of water'.[31] At the international level, the focus on the pricing of scarce resources had pertained mostly to agricultural water; it was but a small step to link pricing to urban water supplies. One element of this transition involved separating the 'public good' characteristics of water (a resource of benefit to all) from its 'private good' characteristics.

Perth's Metropolitan Water Board was the first Australian utility to follow this approach, implementing a system of user pays water rating for the domestic sector on 1 July 1978.[32] Perth's new system stipulated a fixed charge for households, an allowance of 150 cubic metres per annum, and a further charge for each kilolitre used in excess of this amount.[33] In the past, property values had determined the household's water rates, which were independent of the quantity of water consumed. Under the new system, all households had the same 150-cubic metre allowance, regardless of house value, family size, or lot size.[34]

Although there appeared to be general support for the new scheme, some aspects of the system raised the ire of householders, not least its retrospective application.[35] Despite this initial controversy, the new pricing system combined with the social effects of the water restrictions and publicity campaigns rendered household practices' impact on water consumption newly visible.[36] When restrictions were lifted in 1979, a major increase in scheme water use did not occur. The Board's 1985 *Domestic Water Use Study* of consumer attitudes and behaviours regarding scheme water – itself a reflection of growing interest in demand management – observed that the economic pressures of 'pay for use' rating accounted for most of this change in behaviour, particularly among

31 Falkenmark, 'UN Water', 222–27. For an in-depth discussion of the role of this meeting in the formation of a scarcity approach to water management, see Schmidt, *Water*, 115–40.

32 Syme et al., 'Water resources and the quarter-acre block', 46–55. Newcastle and the Hunter District in New South Wales followed two years later in a more widely cited shift to pricing water.

33 MWA, *Domestic Water Use in Perth*, 52.

34 Australian Water Association, *50 years of Water in Australia*, 15.

35 Editorial, 'Think again', *West Australian*, 30 August 1978.

36 See Morgan, 'A thirsty city', 81–97.

non-bore users. Furthermore, the study found that this reduction in scheme water use had most likely occurred outside the home, as indoor water use was very insensitive to price changes. In addition, the study established that in contrast to the mid-1970s, over half of the average household's scheme water consumption occurred inside the home in the mid-1980s.[37]

The shift to user pays reinforced the Water Board's publicity of the importance of water conservation, as household water rates now reflected more closely the costs of providing this service. As for the people of Perth, the new pricing policy recast their identities as consumers. Before, they had been citizens with a right to water, and the government determined their ability to pay for this water through the valuation of their property, which provided a proxy measure of the wealth and income of the consumer and their 'capacity to pay'. Now, under user pays, they were customers, who could purchase different quantities of the commodity, water, according to their willingness to pay.[38] This changing consumer identity resembled the process underway in England and Wales following the nationalisation of the water industry in 1974, with pricing policies developed to ensure the recovery of the cost of supplies. In Perth, the process was a means to manage demand by linking price to consumption, thereby alerting water users to the costs of convenience that networked supplies afforded them.[39] Although the introduction of user pays in Perth helped to reduce water consumption in the late 1970s, some households wore the impact of this reduction more heavily than others.

The loudest opposition to the changes had come from Perth's keen gardeners, who considered that the new rating system discriminated against them. They believed the state government had 'unmercifully slug[ged] thousands of people' who did 'an environmental job in keeping Perth pleasant and presentable'.[40] One asked: 'Is our choice only to be brown lawns, wood chips or paving bricks?'[41] For those fortunate enough to be financially and geographically well situated, private bores and wells could be installed, giving access to virtually unlimited water for garden use.[42] Although private bores have a long history in Perth's suburbs, bore ownership trebled between 1976 and 1982. Bore users

[37] MWA, *Domestic Water Use*, 2, 13, 59, 66. [38] Sharp, 'Water demand', 872.
[39] Bakker, 'Paying for water', 143–64.
[40] B. Reid, 'The water charges called a "slug"', *West Australian*, 10 May 1978.
[41] B. McCartney, 'Quota of water', *West Australian*, 29 August 1978.
[42] Syme and Kantola, 'Investment', 461.

consumed significantly less scheme water than non-bore users at the end of the period, but the *Domestic Water Use Study* estimated that the volume of groundwater extracted by backyard bores was over seven times the amount used for irrigation by non-bore owners.[43] This finding suggested that private bores had been installed so that householders could continue (or even extend) their existing consumption while reducing their scheme water use. This trend persisted into the 1980s, when the Water Board estimated that up to 45 per cent of total water used for domestic consumption in Perth came from private bores.[44] Many consumers therefore interpreted the Water Board's efforts to encourage water conservation as relating only to *scheme* water use rather than to water in general.[45] Although this shift was a predictable result of user pays, its environmental toll on groundwater reserves is still yet to be fully appreciated.

DROUGHT AND DELUGE

In most of the other Australian cities, the 1970s had been a decade of relatively good rainfall with little emphasis on water conservation. 'Australia's metropolitan areas are riding on a wave of water euphoria', the *Australian Women's Weekly* announced in mid-1982.[46] In Melbourne, domestic consumption soared as industrial use stabilised, partly as a result of deindustrialisation and the growing economic importance of services over (sometimes) water-intensive manufacturing. Another reason for the manufacturing sector's more modest consumption was that the Board of Works' increased Trade Waste Charges after the drought of 1967–68, which made it three times more expensive for businesses to dispose of water than to obtain it. As a result, recycling water was a more attractive proposition for industry.[47] The Board of Works provided examples of this approach: in the food-canning industry, water used to wash cans was then used to cool machinery, while at Footscray's Smorgon Consolidated Industries, 'lightly contaminated water' from abattoirs was used in the production of pulp and paper products.[48]

43 MWA, *Domestic Water Use*, 40, 63.
44 Armanasco, 'State position paper', 78.
45 Syme, 'Summary', 160.
46 'Are we using too much water in our cities?', *Australian Women's Weekly*, 4 August 1982, 10.
47 See for example I. Little, 'Letter to the Editor: Water rate unfair to industry', *Age*, 17 July 1968.
48 'Industry offers little conservation scope', *Age*, 24 November 1982.

Carlton and United Breweries' Abbotsford factory reduced the amount of water used per unit of production by 40 per cent and made considerable savings in water and trade water charges as a result.[49] Outside Melbourne's homes, water use rose by over 50 per cent during the 1970s, as Melburnians lavished water on their gardens and driveways.[50] But the 'water euphoria' would not last.

The 1982–83 El Niño drought, most commonly associated with dust storms and the catastrophic Ash Wednesday bushfires, led water utilities to campaign for the households of suburban Sydney, Melbourne, and Adelaide to exercise restraint. Thanks to the additional capacity afforded by the recently completed Shoalhaven scheme, Sydney's Board could opt for a public education campaign to promote the conservation message: 'It's your turn – to save water.'[51] Adelaide's water conservation initiative was the latest in a series of campaigns that had commenced in 1976. The South Australian Engineering and Water Supply Department had encouraged residents to 'Turn It Off', which subsequently become an annual tradition that helped to precipitate a decline in household water consumption by the early 1980s.[52] Such were the drought conditions in 1982–83, however, that the E&WS urged Adelaide households to 'take it real easy this summer' and welcomed the raising of the dam wall at Kangaroo Creek Reservoir.[53]

Rainfall over Melbourne's catchments was just half the long-term average, and rainfall between July and November of 1982 was the lowest on record.[54] After a brief campaign for voluntary restraint, Melbourne households were instructed to cease the use of their sprinklers from December of 1982 to May 1983, with limits on their use of handheld hoses for watering lawns, filling swimming pools, and spraying paved areas.[55] For Melbourne's water managers, the 1982–83 drought had a silver lining: it had made for 'ideal construction weather' for the much-anticipated Thomson Dam.[56] Proposed in 1966 in response to the earlier *Report on the Future Water Supply of the Melbourne Metropolitan Area*, the Board of Works had claimed the dam would make Melbourne

[49] Langford and Heeps, 'A demand strategy', 189.
[50] Melbourne, *Annual Report 1982/83*, 1984, 11.
[51] Sydney, *Annual Report 1982/83*, 13, 16.
[52] *Trends in Water Consumption of Metropolitan Adelaide, 1969–1983*, 1984, 25.
[53] *Trends in Water Consumption of Metropolitan Adelaide, 1969–1983*, 1984, 39–40.
[54] Melbourne, *Annual Report 1982–83*, 1984, 3, 8.
[55] Melbourne, *Annual Report 1982–83*, 1984, 8.
[56] Melbourne, *Annual Report 1982–83*, 1984, 8.

'drought-proof', while Premier Dick Hamer declared its nation-building significance as a 'little Snowy River scheme'.[57] Once completed in 1984, the reservoir on the Thomson River more than doubled the city's water storage capacity.[58]

By the time the Thomson's construction was complete, per capita water consumption in Melbourne had begun to decline, thanks to the greater attention to the role of demand management since the dam's construction began.[59] After formal water restrictions were lifted in December 1983, the MMBW invested in an ongoing advertising campaign, first with the slogan 'We can never take water for granted again' and then the more successful campaign, 'Don't be a Wally with water', featuring the bumbling character Wally, the water waster, and his more water-wise neighbour, Baz (Fig. 7.1).[60] Following the example set in Perth nearly a decade earlier, the Board implemented a user pays system in 1987, which applied an increasing block pricing structure to reward water conservation. Just as Perth had found, this strategy resulted in a sustained reduction of 16 per cent of pre-drought consumption.[61] The success of these initiatives was not without drawbacks. Although these efforts curbed what the Melbourne *Age* newspaper described as the 'galloping consumption' of the suburbs, the Board regretted that the reduced water use had contributed to its reduced income.[62]

Like the other capital cities, Brisbane's water consumption had also escalated, having doubled since sprinklers were permitted in suburban gardens in the mid-1960s.[63] During 1982, the *Australian Women's Weekly* anticipated 'there is likely to be a usage boom, as residents, encouraged to "green the city" for the [1982] Commonwealth Games, leave their sprinklers running 24 hours a day, three days a week'.[64] Events during the 1970s only reinforced Brisbane's sense of plenty. During the austral summer of 1973–74, record-breaking La Niña rainfall across the

[57] Ronalds, *Future Water*; John Jost, '$99m plan to tap Thomson', *Age*, 7 December 1966; "Little Snowy' makes a splash', *Age*, 23 November 1978.

[58] O'Hanlon and Spearritt, 'From water engineers', 102.

[59] Viggers et al., *Melbourne's Water*. 89. See Langford et al., 'A demand'; Jackson et al., *A Water Plan*.

[60] Bronwyn Cran, 'Comic ads to promote water conservation', *Age*, 12 January 1985.

[61] Duncan, 'Water demand', 26.

[62] Tim Colebatch, Carol Sides, and Richard Guilliatt, 'A problem of galloping consumption', *Age*, 24 November 1982; Melbourne, *Annual Report 1985–86*, 1986, 15–16.

[63] Brady and Cossins, 'Factors'.

[64] 'Are we using too much water in our cities?', *Australian Women's Weekly*, 4 August 1982, 10.

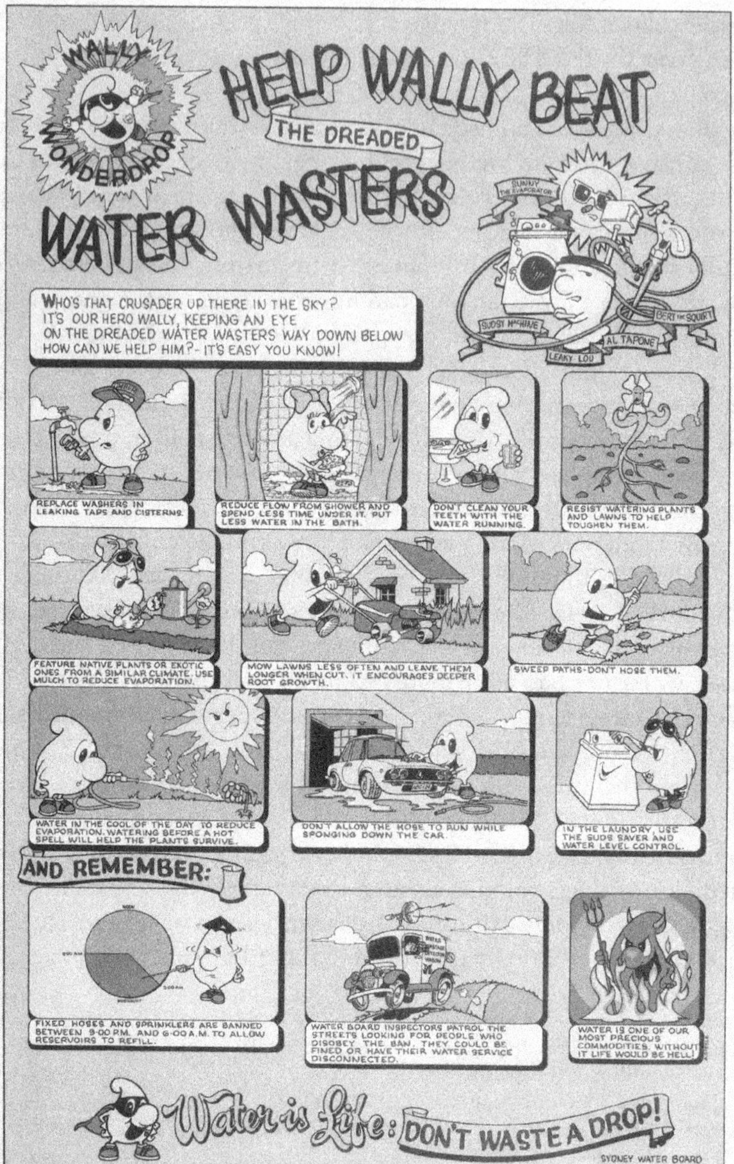

FIGURE 7.1. *Water is Life*. In the 1980s, Sydney Water Board employed 'Wally Wonderdrop', among other means, to encourage water-conserving behaviour among residents.

FIGURE 7.2. Edward Street, Brisbane, during the 1974 floods (State Library of Queensland).

state brought one of the wettest summers in many years. As the city prepared for the Australia Day public holiday on 26 January, a large monsoonal trough, associated with Cyclone Wanda, hovered above the Brisbane River catchment, an area similar in size to twenty-first-century Tokyo. On 26 January alone, the city received in excess of its average monthly rainfall.[65] On 29 January, the flooding peaked at 5.45 m at the Brisbane GPO gauge, the highest since the level of 8.34 m in 1893. Awe-struck spectators watched as the turbulent waters raced at nearly 25 km an hour towards Moreton Bay.

Local newspapers, the *Courier-Mail* and *Sunday Mail*, reported 'houses were swept into raging floodwaters', while, 'ripped off their stumps, steel walls of factories were torn open, and luxury craft were smashed to matchwood'.[66] The Brisbane River – now more than 3 km across at its widest, having swollen into its floodplain – submerged or destroyed everything in its path. Floodwaters cut across highways and railroads and closed the airport: only one bridge over the river remained open, as streets became canals. Before long, Brisbane was largely isolated. Flooded suburbs were left without gas and electricity supplies, and raw sewage poured into the river due to inundation of the city's energy and

[65] Director of Meteorology, *Brisbane Floods*, 29–31.
[66] Cook and Spearritt, 'Water forever'.

sanitary infrastructure. The damage extended upstream, with land and buildings submerged up to 9 m deep; coal production at nearby Ipswich ground to a halt.

Fourteen people died in the Brisbane floods; 13,000 homes across thirty suburbs were left either submerged, inundated, or otherwise damaged. People wept openly in the streets as they returned to what remained of their homes to find their possessions ruined or their homes damaged beyond repair. With agricultural land rendered unproductive and food left rotting in the Brisbane Markets, the community faced food shortages. Numerous industries, shops, and businesses struggled to resume work, threatening the viability of businesses and the livelihoods of their employees.

Despite calls for an inquiry into the flood, the conservative National Party State government, under Premier Joh Bjelke-Petersen, was adamant that such an inquiry would be 'pointless' as the cause – heavy rain – was clear.[67] According to this logic, averting similar disasters in the future was a matter of further hard engineering intervention. The deluge expedited the State government's plans to construct the Wivenhoe Dam, a flood mitigation dam on the Brisbane River that had first been mooted in 1971. Such was the prevailing faith in Queensland's dams, there was no state policy on flooding nor regulation of floodplain development, unlike New South Wales and Victoria. Outside the Bjelke-Peterson government, questions remained as to the contributing role that land laws and development policies had played in the disaster. The Union of Australian Women, for instance, demanded a full public inquiry: 'We are convinced, like many other Queenslanders, that too many decisions about so-called development are made in the interest of profit for the "developers" rather than improvement of the quality of the lives of the people and the security and safety of their homes.'[68]

A decade after the floods, the Wivenhoe Dam was opened in 1985, some 80 km upstream on the Brisbane River. For long-serving Premier Bjelke-Petersen, it was one of the triumphs of his government (1968–87). For the residents of Brisbane, the dam's construction created a 'double illusion'.[69] It fostered not only an illusion of security from flooding but also a sense of endless water supplies for the city. This outlook was a welcome relief after the six months of water restrictions imposed on Brisbane's households during 1978, during which time the use of

[67] Cited in Cook, 'Vacating the floodplain', 346.
[68] Cited in Cook, 'Vacating the floodplain', 346.
[69] Head, 'Managing urban'.

'unattended garden sprinklers' was banned.[70] Otherwise, there had been few incentives for Brisbanites to curb their water use, as both major political parties in the Brisbane City Council agreed to halt the installation of water meters in the 1980s. The proportion of houses with meters fell from 80 per cent in 1930s to just 6 per cent, leaving most residents unaware of their true water usage.[71] Brisbane residents were simply rated on the unimproved capital value of the land they occupied; the Chief Engineer and Manager of the Brisbane City Council's Department of Water Supply and Sewerage estimated that the city's households paid more for a small can of Coca Cola than for their water each day.[72] Brisbane did not introduce a two-part tariff for residential water, with the price increasing as more water was used, until 1997, over a decade after Melbourne and Sydney had opted for price signals as one way of curbing suburban water demand.[73]

POLLUTION AND PROTECTION

In his response to the 1978 McKinsey report into the operations of the MWSDB, the union secretary warned that without adequate funding, 'our beaches, rivers and streams will continue to be polluted for the foreseeable future'.[74] This appeal to wider concerns about the degraded condition of Sydney's waterways suggests that the situation had become intolerable. A Randwick correspondent to the *Sydney Morning Herald* described the city's ocean sewage outfalls as 'little short of an assault', while another observed, 'what you flush away today may cling to you in the surf tomorrow'.[75] These were problems shared by all Australia's major cities to varying degrees, as the 1970 Senate Select Committee on Water Pollution had found. Melbourne's Yarra was coated with a sickly sweet slick thanks to discharge from a nearby confectionary factory.[76] Scientists identified the Georges River in Sydney as the source of 2000 cases of oyster-associated

70 *Brisbane City Council Annual Report 1978–79*, 22.
71 Cole, *Shaping a City*; Spearritt, 'The water crisis', 20.
72 See Cole, *Shaping a City*. He estimated that the daily household cost of water was 18 cents, while the price of a can of Coca Cola was 22 cents; O'Connell, 'Major city', 15.
73 O'Hanlon and Spearritt, 'From water engineers', 103.
74 Bob Hunt, 'The disposal of sewage', *Sydney Morning Herald*, 4 January 1979.
75 Kenneth P. Finn, 'Discharging waste into the ocean', *Sydney Morning Herald*, 2 January 1979. David Dale, 'Cities choke in wastes that won't go away', *Sydney Morning Herald*, 19 July 1971.
76 Ted Baillieu, 'Melbourne (Flinders Street Land) Bill Second Reading', *Legislative Assembly*, 30 April 2003, 1192.

gastroenteritis in 1978, courtesy of sewage contamination.[77] Foam gathered several feet deep in numerous small streams in the Brisbane area, due to non-biodegradable detergents.[78] 'Cholera type' organisms were found in Brisbane drinking water in 1977.[79] Even the heir to the British throne had fallen foul of Port Phillip Bay, which Prince Charles had reportedly described as 'like swimming in diluted sewage', after a visit in 1970.[80]

The sights, smells, and sounds of the Australian suburbs were on the eve of change; the use of suburban waterways as drains for industrial waste would be tolerated no more. The ecological awakening of the 1960s, heralded by the publication of Rachel Carson's *Silent Spring* (1962), reflected a growing international concern for the environment and the toll human activity was taking upon it, as well as the implications of environmental degradation for human health. Writing for lay readers, Carson had clearly and evocatively explained the harmful impacts of the misuses of industrial pesticides. Among her case studies, she cited evidence from patients at Prince Henry's Hospital in Melbourne that supported her assessment of the relationship between prolonged exposure to insecticides and mental health.[81]

Carson's work was well received in Australia. The official journal of the Australian Medical Association, the nation's peak professional body for medical doctors, observed in 1964 that despite its 'emotional overtones', it had 'highlighted the problems of abuse and recklessness in the employment of pesticides. Moreover, it has taught the man in the street a new word – ecology – and has made him aware of its connection with man's ultimate welfare in the world'.[82] Carson's writing also resonated with Australian women. In response to a review of Australian ornithologist Jock Marshall's 1966 collection, *The Great Extermination: A Guide to Anglo-Australian Cupidity, Wickedness and Waste*, Louisa Patton of Melbourne wrote to the *Age* to remind its readers that women too 'are alert to the menace described'. She reported a resolution that had passed unanimously at the recent 78th annual convention of the Women's Christian Temperance Union of Victoria:

> We urge that immediate action be taken to control the use of chemical pesticides which are contaminating areas of our land and waterways with dangerous,

77 Reid and Bone, 'The rise'.
78 *Report from the Senate Select Committee on Water Pollution* (1970), 22.
79 'Cholera scare strikes', *Age*, 9 April 1977. 80 Dale, 'Cities choke'.
81 John Hetherington, 'Nature jeopardised by pest poisons?' *Age*, 13 April 1963.
82 'Pesticides', *Medical Journal of Australia*, 11 January 1964, 49.

poisonous materials; and that adequate research and testing be assured for the discovery of alternative means of eradicating plant, animal or other pests before or instead of using what is now acknowledged to be lethal.

This resolution, Patton noted, was based on material from the UN Food and Agriculture Organisation and Carson's *Silent Spring* – 'significantly, a woman scientist', she added.[83]

Pesticides were among the contaminants discussed by the Senate Select Committee on Water Pollution, chaired by South Australian Liberal senator Gordon Davidson. Established in 1968, shortly after a fire at a chemical factory released poisonous parathion gas (a carcinogenic insecticide also used in chemical warfare) into the Brisbane River, the Committee took evidence from governments, utilities, businesses, universities, and non-government organisations, such as the Queensland Littoral Society and the Mount Lofty Ranges Association in South Australia.[84] Its report, popularised by journalist Don Whitington's *The Effluent Society*, revealed that Australians, far from living in an environmental paradise, were perhaps, per capita, some of the worst environmental offenders.[85] Having raised urban water supply and waste disposal to world-leading standards by 1900, it was as if Australian governments had grown complacent – that, having reduced major health risks, little more needed to be done. Such rising ecological and human health concerns, and the campaigns that working- and middle-class activists devised to amplify them, encouraged State governments to adopt a suite of air, noise, and water pollution regulations.[86] Modelled on US legislation, these regulations included the *Clean Waters Act 1970* in New South Wales and the *Environment Protection Act 1970* in Victoria.

The passage of such legislation reflected concerns about not only pollution but also the impacts of the nation's faith in developmentalism. Although the continent's unique flora and fauna had long captured the attention of nature enthusiasts, the post-war environmental movement recognised that such species belonged to an interdependent system of life that required conservation and protection as a whole.[87] Preventing the destruction of these areas became the focus of the growing Australian conservation movement, as it campaigned to place the environment on

83 Louisa Patton, 'Gradual extermination', *Age*, 31 May 1966.
84 *Report from the Senate Select Committee on Water Pollution* (1970), 24.
85 Whitington, *The Effluent Society*.
86 See for example Goodall and Cadzow, 'The people's'.
87 Bonyhady, *The Colonial Earth*.

the political agenda and to politically legitimise its concerns.[88] Taking shape were groups such as the Australian Conservation Foundation (1965) and the Total Environment Centre (1972), while Australian conservationists formed the world's first 'Green' party, the United Tasmania Group, in early 1972, in the wake of the damming of Lake Pedder in Tasmania. When the Tasmanian government later decided to dam the Franklin River, this group joined with the Wilderness Society and others to protest the proposal, culminating in the Franklin Dam blockade in late 1982. Support from these groups soon proved decisive for the Australian Labor Party at the 1983 Federal election, after which the Hawke government introduced new laws to protect the Franklin River.[89] This movement became the foundation for a national Australian Greens party, which formed in 1992 under the leadership of Franklin Dam activist and Tasmanian parliamentarian Bob Brown.

Similar concerns extended to the environmental impacts of water resource development projects for the five cities. By the mid-1970s, local environmentalists had joined in an unlikely coalition with opponents to construction of the Thomson Dam, along with Gippsland irrigators and advocates of industrial and coal development in the Latrobe Valley, who all feared that the loss of water to Melbourne would hinder their interests.[90] The Hamer government was confident, nonetheless, that the dam aligned with its decentralisation policies by providing more water for Melbourne, while regulating the river's flow to the advantage of downstream users in Gippsland, as well as local businesses and subcontractors.[91] Under the State government's nascent program of environmental impact assessment, the project was subjected to scrutiny, although the Conservation Council of Victoria was wary that the government would proceed regardless of the report's findings, given its enthusiasm for the dam.[92] The Council warned of the dam's potential to affect the Gippsland Lakes as 'wetlands, waterfowl habitat, fishing grounds and watersport areas' and suggested there were alternative sources to ensure Melbourne would be 'drought-resistant'.[93] Published in September 1975, the environmental impact study claimed, however, that the dam would do no significant damage to its environment or the waters downstream,

[88] Bonyhady, *Places Worth Keeping*. [89] Hutton and Connors, *A History*.
[90] Tim Colebatch, 'Dam to treble our water supply', *Age*, 26 December 1975.
[91] 'Coal, water, oil, timber, paper', *Age*, 14 May 1978.
[92] Ministry of Conservation, *Assessment*; MMBW, *Report on Environmental Study into Thomson Dam*.
[93] 'Delay start on river dam, says Council', *Age*, 29 November 1975.

despite resulting in the diversion of almost three-quarters of the Thomson's annual flow from the Lakes to Melbourne's water supply.[94] The study had not considered the effect of the dam on the future water demands of the Latrobe Valley and conceded that there was insufficient data to estimate its effects on the Lakes. Environmental concerns about the dam's impact continued after its completion in 1984, not least as the Lakes had been declared a Ramsar site in the interim.[95] Over a decade later, the *Age* described the Lakes as in a 'severely degraded' condition as a result of 'farm pollutants and low river inflows due to high diversions for irrigation and the Thomson Dam'.[96]

In the meantime, a potassium cyanide spill on Sydney's Cooks River, which drains into Botany Bay, had tested the new regime of the New South Wales *Clean Waters Act*. By the early 1970s, the Cooks River was the recipient of about a quarter of the liquid wastes from Sydney's factories.[97] In December 1973, around 55,000 litres of chemicals leaked from the Sunbeam Corporation's plating plant, resulting in the deaths of thousands of fish, birds, and eels.[98] Fifteen tons of dead fish alone were collected from the river in the aftermath.[99] The incident precipitated the first water pollution charge in the state, where the Supreme Court fined the company \$3000 (of a possible \$10,000) for the accident.[100] Although the fine was relatively modest, the ruling set a precedent whereby a business was subject to the provisions of the *Clean Waters Act* even if the pollution was accidental.[101] Subsequently, the waterway became the first in the state to be brought under the regulation of the State Pollution Control Commission in 1975, which classified the Cooks as 'Restricted Waters', whereby discharged waters should be of quality adequate to sustain aquatic life.[102]

94 Colebatch, 'Dam to treble our water supply'.

95 Robert Lawson, 'Freshwater flow cut is threat to Lakes system', *Age*, October 1986; Andrew McCutcheon, 'Water flow to Gippsland Lakes to be maintained', *Age*, 16 October 1986; Leith Young, 'More water for city leads to saline lakes, says Hunt', *Age*, 20 October 1986; Robert Lawson, 'Open Thomson', *Age*, 23 October 1986; Andrew McCutcheon, 'Lakes salinity', *Age*, 4 November 1986; Meredith Hayes, 'Gippsland Lakes pollution was predictable', *Age*, 20 January 1988.

96 Claire Miller, 'Pesticide spray plan draws fire', *Age*, 18 October 2000.

97 Joy and Hickson, 'Of sewers and drains', 161.

98 'Chemical kills thousands of fish, birds', *Sydney Morning Herald*, 28 December 1973.

99 Tyrrell, *River Dreams*, 171.

100 Tyrrell, *River Dreams*, 172; '\$3,000 penalty: cyanide in river', *Sydney Morning Herald*, 28 June 1974.

101 Joy and Hickson, 'Of sewers and drains', 180.

102 Coward, 'Reorganising waste', 44.

By the mid-1970s, the quality of some of Sydney's waterways had begun to improve, particularly the Duck and Cooks rivers and the Alexandra Canal. Another beneficiary of the emergent regulatory regime was the Parramatta River, upstream parts of which (near the future site of the 2000 Olympic Games) had been declared 'biologically dead' in the mid-1960s.[103] Although some commentators still described the river as an 'ecological disaster', the *Sydney Morning Herald* celebrated the return of fish species such as mullet a decade after the introduction of the *Clean Waters Act*.[104] Another factor in the Parramatta's recovery was the intervention of the state's National Trust, which advocated for its restoration with a strategic plan to strengthen pollution controls, clean the banks of debris, landscape industrial lands, and acquire additional riverside parks.[105]

It was on the banks of the Parramatta that the State's first Green Ban took place in 1971. The Builders Labourers Federation joined a group of local middle-class women (the 'Battlers for Kelly's Bush') to prevent bushland from being cleared for luxury homes.[106] By the mid-1970s, over forty Green Bans had been declared in Sydney, as unionists joined with local residents to oppose large-scale redevelopment projects.[107] The Green Bans attracted wide international interest, such was the unprecedented nature of unionists withdrawing their labour to help protect low-income housing areas, historic buildings, and bushland. Visiting West German activist Petra Kelly reportedly took the concept back to Europe, while Paul Ehrlich described them as 'the most exciting ecological happening, not only for Australia but overseas as well'.[108] For Ehrlich, the Bans had reportedly represented the 'birth of urban environmentalism'.[109]

Over a decade after a Green Ban had helped Indigenous activists preserve Sydney's Redfern, in 1989 members of the Construction, Mining, and Energy Workers' Union in Perth withdrew their labour from the Old Swan Brewery site at the foot of Mount Eliza on the Swan River.[110] At the sacred site of Goonininup, Whadjuk Noongar had been

[103] 'Good fish catches in the Parramatta', *Sydney Morning Herald*, 18 January 1976.

[104] Nancy Berryman, 'A river fights for its former grandeur', *Sydney Morning Herald*, 8 March 1981; Barnaby, 'Australia', 27–33.

[105] Berryman, 'A river fights'; Lynch and Pearson, *Parameters*.

[106] Berryman, 'A river fights'. [107] Freestone, 'Australian environmental', 26.

[108] Cited in Mulligan and Hill, *Ecological Pioneers*, 269; Burgmann and Burgmann, *Green Bans*, 9.

[109] Cited in Mulligan and Hill, *Ecological Pioneers*, 269.

[110] On Redfern, see Cook and Goodall, *Making Change*, 25–60; Anderson, 'Place narratives', 314–35.

campaigning to stop its redevelopment into a luxury apartment complex, which was slated to be another jewel in the boom State's glittering crown.[111] For Whadjuk Noongar, Goonininup is the sacred ground of the Waugal, an area that the Director of the Aboriginal Medical Service described in a 1989 statement of support as

> a site where the Waugal rested and drank from the spring which was created to ensure there would always be plentiful water, a site which would always be special for the indigenous people that lived there. It is a site to which the Waugal continually returned after tiring of creating the Swan and Avon rivers before continuing on to create features like the Darling Range and Murray Rivers. This is the site we want protected. The Waugal has always looked after the children of the Swan River – it is now our turn to look after Waugal. Waugal has to have a resting place.[112]

To the Fringe Dwellers of the Swan Valley, a Noongar group, 'if the plans go ahead, it is digging into our grandfather's grave. It is digging into the side of our grandfather's grave'.[113] Having earlier occupied Matagarup (Heirisson Island) on the Swan River to campaign for land rights, this group had recently taken the State Energy Commission to the Supreme Court to oppose the construction of a gas pipeline under Bennett Brook (about 21 km northeast of Goonininup), which they had successfully argued would disturb the Waugal.[114]

While a petition wrapped in paperbark was dispatched unanswered to the Queen, Noongar and their supporters formed a protest camp at the site in January 1989.[115] As one Noongar woman argued, 'the only way we are going to get anything is by protesting', while another summarised her community's anguish:

> This place is significant to all Aboriginal people but especially to the Nyungahs. This is their very special place on the Swan River. How can the Government insult the people yet again by not being bound by their own laws so they can put a Northern Territory art collection in a building which distilled alcohol, a problem to our people from the beginning of white settlement? It's such a small piece of land; yet is so sacred to us!! Let it be parkland, not a profit-spinning sideshow for the Government.[116]

[111] Jacobs, *Edge of Empire*, 103–31.
[112] Ted Wilkes, 18 April 1989, in Ansara, *Always Was*, 71–72.
[113] Robert Bropho and Louis Nettle, 4 May 1988, cited in Ansara, *Always Was*, 65.
[114] 'Pipeline takes a new twist to by-pass sacred serpent', *Sydney Morning Herald*, 16 October 1986.
[115] Ansara, *Always Was*, 37.
[116] Cited in Ansara, *Always Was*, 36.

Joining the diverse coalition of protesters were church groups and heritage advocates, reflecting a growing movement towards social and political justice for Aboriginal people. Nine months later, police forcibly removed the camp, while the Federal Labor government backed down after a brief intervention in the matter. Further attempts to prevent the redevelopment were defeated in the courts, and in mid-1992, the State Labor government controversially gave the site to the private developer Multiplex on a sixty-five-year lease with a peppercorn rent.[117]

SEWAGE AND THE CITY

The growing awareness of the extent of the pollution of suburban waterways had contributed to the increasing popularity of backyard swimming pools. The nation's longest-running magazine, the *Bulletin*, speculated as much, reporting on the 'spectacular' surge in sales of Clark Rubber's above-ground outdoor swimming pools in late 1972.[118] It attributed the company's booming sales in Melbourne to 'its polluted hydrosphere':

> The current scare reports make it quite plain that only those afflicted with a death wish would submit themselves to a baptism in the city's mephitic bayside or its unspeakable affluent, the Yarra. Sydney too, has been busy fouling its own nest, with the result that the near-city beaches are now widely unappreciated.[119]

Once beyond the means of middle- and working-class households, the fully tiled in-ground pool was now competing with recent innovations, such as fibreglass and vinyl-lined swimming pools, as well as the significantly cheaper above-ground option. A Morgan Gallup poll in 1976 found 11 per cent of Australian homes had swimming pools, with nearly half of them in New South Wales.[120]

Although Sydney had benefited significantly from the National Sewerage Program, the destination of the city's waste remained unchanged. In discharging its wastes into the sea, Sydney was not alone among the five cities, but unlike the others, it did not subject its wastes to secondary treatment.[121] Although this discharge was unpalatable to human residents, these wastes were a boon to wildlife, due to the low fertility of Australian soils and coastal waters. Migratory sandpipers and knots feasted in the mudflats north of Brisbane's Luggage Point before

[117] Shaphan Cox et al., 'Indigenous persistence'; Lewi, 'Whose heritage; Gregory, *City of Light*: 269–80.

[118] Myers, 'Swimming pools', 63–64.

[119] Myers, 'Swimming pools', 63–64.

[120] Myers, 'Swimming pools', 63–64.

[121] Senate Select Committee, 1970, 50–51.

flying north to Siberia. Until the 1980s, Sydney's cliff-face outfalls likewise sustained vast numbers of the now-endangered wandering albatross, which was especially partial to the refuse of the abattoirs at Homebush, the future site of the Olympic Games. Banding studies indicated that the albatrosses had travelled from their breeding colonies as far away as the islands off the east coast of South America.[122] Such was the popularity of Sydney's waters to the likes of southern skuas, black-browed albatrosses, and northern and southern giant petrels, the Australian biologist Tim Low could marvel in 2002: 'No other city below the equator fed so many big birds ... Sydney was the ultimate food trough.'[123]

In Melbourne, the treatment plant at Werribee on Port Phillip Bay also proved popular with local and migratory shorebirds of the East Asian–Australasian Flyway. Werribee's unintentional appeal to diverse animal, plant, and bird life had been recognised since the early 1920s, and in 1982 the entire area was declared a wetland of international significance under the Ramsar convention, with bird watchers recording over 250 bird species there, including the critically endangered Orange-bellied Parrot.[124]

Plants also thrived on human wastes. The flow of household and industrial wastes into the Georges River fed the voracious appetite of the alligator weed, first recorded in Australia in the 1940s. Rainfall run-off, especially from unsewered areas, flushed nutrients into the river, further encouraging the growth of tangled mats of weed across the water's surface at the expense of other aquatic species.[125] By the early 1970s, the alligator weed had flourished to such an extent that it covered about 18 km of the upper reaches of the river, and heavy rains washed large 'rafts' of the 'creeping death' into Botany Bay.[126] Having observed the success of the US Department of Agriculture in combating the pest, the State Pollution Control Commission approved the introduction of the alligator weed flea beetle in 1976 and a moth (*Vogtia malloi*) the following year. These creatures quickly curbed the weed's growth, only for duckweed and then Salvinia to take hold. Although spraying the latter with chemicals such as hexazinone, paraquat, and glyphosphate had some success, CSIRO scientists noted in the early 1980s that 'the

[122] Low, *The New Nature*, 61. [123] Low, *The New Nature*, 62–63.
[124] Penrose, *Werribee Farm*, 41.
[125] Sainty, McCorkelle, and Julien, 'Control and spread', 195–201.
[126] Gavin Souter, 'At war with the triffids', *Sydney Morning Herald*, 1 December 1973; 'Control of alligator weed study begins', *Sydney Morning Herald*, 4 June 1974; 'Creeping death', *Sydney Morning Herald*, 22 May 1974.

problems are likely to continue as long as the effluent from the [Glenfield] sewage treatment plant is allowed to discharge into the river'.[127]

The Commonwealth Local Government Development Program could attribute the dwindling popularity of public swimming pools to Australians' 'increased demand for home-based activities'.[128] Over the following decade, the legislatures that managed these facilities attempted to respond to these changing expectations with visions of the 'Pool 2000', to borrow from a 1992 Brisbane report.[129] Among the pools slated for closure and replacement as a result was Melbourne's Fitzroy Municipal Baths, built in 1908 when sanitary facilities were still rudimentary for many households. In October 1994, the Baths became one of the targets of the neoliberal reform agenda of the Victorian Liberal Kennett government, which deemed its meagre returns financially unsustainable. After petitions and protests to 'Stop the Slash, Maintain our Splash!' as well as Commonwealth intervention and an eight-week occupation, the Baths earned a reprieve.[130]

Accompanying this shift towards neoliberal public policy was the adoption of the international agenda for sustainable development. The Federal Labor government's *National Strategy for Ecologically Sustainable Development*, the Australian response to the Rio Earth Summit's Agenda 21, emphasised the 'need to develop ways of using those environmental resources which form the basis of our economy in a way which maintains and, where possible, improves their range, variety and quality'.[131] Announced in 1992 in concert with the more aggressively pursued *National Competition Policy*, the sustainable development strategy shared its language of efficiency and market valuation while advocating the need for the 'participation of every Australian'.[132] Among its programs was Waterwatch, which followed the established Landcare model of rural volunteerism for the amelioration of local soil problems, such as salinity and erosion. The umbrella Waterwatch program oversaw a range of state-based programs, such as Rivers of Blue in Western

[127] Finlayson and Mitchell, 'Management of salvinia', 71–76; Walters et al., 'Different engineering', 52–59.

[128] Robertson and Keady, *Holroyd Municipal Swimming Centres*, 12.

[129] Wood, *Brisbane City*, 28.

[130] Benjamin Preiss, 'Community groups swim against the tide in a bid to save outdoor pools', *Age*, 4 February 2015.

[131] Ecologically Sustainable Development Steering Committee, *National Strategy for Ecologically Sustainable Development.*

[132] Dovers, 'The rise', 39.

Australia and Streamwatch in New South Wales, in which volunteer groups and schools undertook environmental education and the monitoring of local waterways. Critics of its implementation pointed to the state's instrumentalisation of local concerns about the health of suburban (and rural) waterways.[133] Devolving their responsibilities for water management to 'community groups that have a particular interest in, or capacity to contribute to ESD', as advocated in the *National Strategy*, was a means that critics considered served to maximise revenue and reduce spending.[134]

Sydney residents and beach-goers would no longer tolerate the city's inadequate sewage infrastructure. The city's rivers were getting cleaner, but as researchers had observed in 1975, 'the problems are being modified and shifted from river-dwellers and river-users to beaches and beach-users'.[135] A decade later, beach-goers could encounter all manner of grease, faeces, condoms, tampons, and maggots, as southerly and south-easterly winds blew the effluent back to the shore.[136] What was not visible to the naked eye was perhaps even more worrying. While the authorities held that 2000 parts per 100 ml were 'acceptable' levels of faecal coliform, reported levels at some beaches were as high as 50,000 parts per 100 ml.[137] Many swimmers and critics attributed viruses and skin complaints to their exposure to these conditions, although the Water Board remained tight-lipped on the matter.[138] Another cause for concern was the amount of toxic waste that the outfalls disposed into the ocean. Sharon Beder, an outspoken engineer and author, estimated alarming quantities of heavy metals and chemicals, such as cadmium, chromium, copper, lead, mercury, cyanide, and chlorinated hydrocarbons were flowing into the sea and contaminating marine life.[139]

The Pacific was still serving as a cheap sink for Sydney's sewage, instead of the costlier option of secondary treatment, estimated to be at

133 For critiques, see for example, Dovers, 'Institutions', 17; Summerville, Adkins, and Kendall. 'Community participation', 696–711; Searle, 'In search', 99–109.

134 Ecologically Sustainable Development Steering Committee, *National Strategy*, 10.

135 Joy and Hickson, 'Of sewers', 137.

136 Robin Hill, 'Has Surfline too much dirt on our polluted beaches?' *Sydney Morning Herald*, 24 January 1987.

137 Alan Tate and Paul Bailey, 'Polluted beaches: what parents should know', *Sydney Morning Herald*, 11 January 1989.

138 Neale Prior, 'The crying shame of polluted beaches', *Sydney Morning Herald*, 29 September 1988.

139 Frank Walker, 'Toxic waste still pouring into sea', *Sun Herald*, 29 January 1989.

least ten times more expensive than ocean disposal.[140] Without additional treatment, the government's plan to extend the outfalls further into the sea would only have a cosmetic effect on Sydney's beaches, while allowing the ocean's contamination to continue. As Beder observed in 1989, 'the Water Board has the attitude that the ocean is infinite, but it is not. These metals do accumulate.'[141] With fishing soon banned around the outfalls, revelations in the local press combined with the efforts of activist groups to rally concern about the human and environmental impacts of Sydney's sewage.[142] These culminated in the 'Turn Back the Tide' concert on Good Friday 1989 at Sydney's Bondi Beach, where 250,000 people showed up to protest the pollution of their beaches and waterways.[143]

Two years later, the extension of the city's three ocean outfalls was completed, a project that had commenced in 1983 (Map 7.1). No longer discharging effluent from the cliffs over North Head, Malabar, and Bondi, these deepwater outfalls comprise tunnels that are up to 4 km long in the bedrock of the sea floor, with lines of diffusers to disperse the waste into the sea. Out of these tunnels flowed about a billion litres of sewage a day, the equivalent of about 400 Olympic swimming pools.[144] While the government boasted that Sydney's beaches were the 'cleanest they have been for years', critics argued that the problem of the city's sewage would become 'out of sight, out of mind'.[145] For the albatrosses, however, the demise of Sydney's coastal 'river of lard' may be partly responsible for their plummeting numbers, as some birders suspect.[146]

REFORM AND REGRET

The New South Wales government's reluctant concession to extend Sydney's sewage outfalls reflected wider changes underway in the organisation of Australia's urban and rural water utilities. From the early 1990s, Conservative governments, particularly in New South Wales and Victoria, wanted to curb the power and the apolitical expertise of

140 Beder, 'Getting into deep water', 62–74; Coward, *Out of Sight*, 298.

141 Cited in Frank Walker, 'Toxic waste still pouring into sea', *Sun Herald*, 29 January 1989, 39. See also Beder, *Toxic Fish*.

142 Bernard Lagan and Andrew Keenan, 'Fishing banned near sewage outfalls', *Sydney Morning Herald*, 24 February 1989.

143 Hawkins, 'Shit in public', 31–32.

144 Ben Hills, 'Sewer City', *Sydney Morning Herald*, 28 December 1991.

145 For example, Joe Schipp, 'Boats and sewage', *Sydney Morning Herald*, 10 December 1991; Ben Hills, 'Sewer City', *Sydney Morning Herald*, 28 December 1991.

146 Low, *The New Nature*, 60–62; Battam and Smith, 'Report on review'.

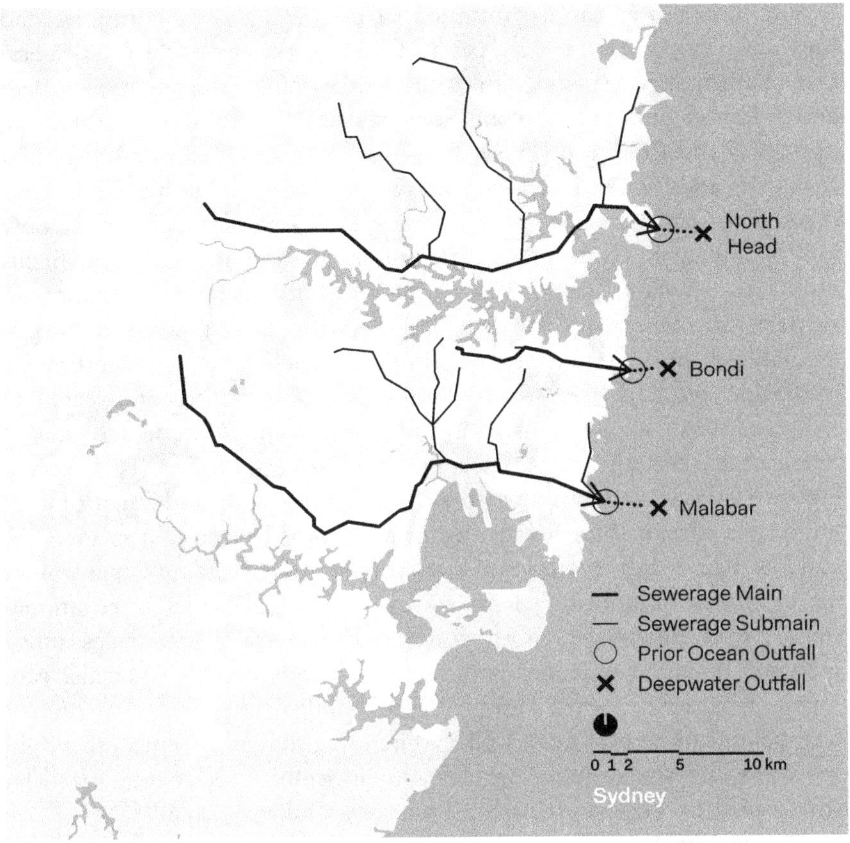

MAP 7.1. Sydney Ocean Outfalls map (drawn by Daniel Jan Martin).

engineers that had made the MWSDB and MMBW successful and powerful planning and utility bodies. The Greiner Liberal government abolished the MWSDB in 1994, cutting back on staff and demanding a better return on capital, so that the new body retained less of its earnings, which was how statutory authorities had traditionally funded new works and maintenance. The new Sydney Water Corporation was charged with protecting the environment and public health and operating in a commercial manner by promoting user pays pricing to encourage water conservation. Ecological risk assessments were to be performed in conjunction with the Environmental Protection Agency.[147]

[147] State Records NSW, data sheet; see also Freeman, *Going Corporate*. On the Greiner era, see Sheil, *Water's Fall*, 98–100.

Such changes to the organisation of the nation's water utilities had been underway since the early 1970s. Globally, the end of the 'golden era' saw challenges to Australia's standard of living. Rising unemployment and inflation and falling competitiveness saw both Labor (Whitlam, 1972–75) and Coalition (Fraser, 1975–83) national governments at first cut tariffs and then reduce expenditures and increase flexibility in labour markets to improve economic outcomes. Both sets of policies made only modest impact on the stagflation (high inflation and high unemployment) of the late 1970s. In the USA and UK (both usually influential on Australian policies), the Reagan and Thatcher governments cut taxes, regulations, and government institutions heavily, while significantly increasing social divisions. The Hawke Labor government (elected in 1983) adopted a more consensus-based approach. While it deregulated financial markets and exchange controls, it also established 'The Accord' between labour and business to slow wages growth and inflation. An important part of the package, again influenced by changes in the USA and UK, was a shift to a smaller role for government and an expansion of market-based solutions. These changes, while following international trends, were introduced more slowly than overseas. Nonetheless, they followed the same essential paradigm, as summarised by Marxist economic geographer David Harvey: 'Neoliberalism proposes that human well-being can be best advanced by liberating individual entrepreneurial freedoms and skills within an institutional framework characterized by strong private property rights, free markets, and free trade.'[148]

This neoliberal view that markets, rather than government bureaucracy, were the best way to allocate resources came to dominate Australian water management, as it did elsewhere. By 1992, the 'Dublin Statement' of the International Conference on Water and the Environment listed the recognition of the economic value of water as a core principle for managing environmental and human priorities in the twenty-first century. Water, according to the Dublin Statement, 'has an *economic value* in *all* its competing uses and should be recognised as an economic good. *Within this principle*, it is vital to recognise first the basic right of all human beings to have access to clean water and sanitation at an affordable price.'[149] One simple response to the increasing perception that water was valuable and scarce was to be less wasteful and reduce demand. Resources such as clean water and a natural environment were valuable – and, given increased

[148] Harvey, *A Brief History of Neoliberalism*, 2.
[149] World Meteorological Organization, *The Dublin Statement*.

public awareness, needed to be used more carefully. Allotted a higher price, advocates argued, water would be used with caution. Once citizens were recognised as having a 'right' to a clean environment, those who wished to damage that right needed to pay a price. Identifying property rights facilitated ownership – and the need to either pay (or negotiate) access – whether to clean water or a clean environment. Creating mutually acceptable rules of behaviour could thus reduce costs of misuse and conflict and produce more efficient outcomes.[150] Economists argued the advantages of market-based solutions to signal the 'true cost' of resources to individuals, but for many people, the 'right to water' still meant governments, rather than markets, were the acceptable institution to distribute water.

From the federal government's perspective, microeconomic reform, initially a set of ad hoc policies aimed at increasing productivity and competitiveness, over time became a systematic program that dominated change at State and local government levels by the early 1990s.[151] Although the introduction of competition principles was not uncontested, soon many areas of Australian life, including trade, labour markets, education, and health care, became subject to the renewed focus on efficiency. Water resource management was one of many fields where government provision and regulation, previously seen as the only method of coordination, were to be replaced by a more market-based approach.

Water had long been associated with national development, while clean water was vital for public health. As noted in Chapter 4, in both situations the public good aspects aligned well with public provision, especially when the high initial costs of water infrastructure were prohibitive to private provision and economies of scale made water supply and waste removal natural monopolies. By the 1980s, rising concerns over water scarcity and quality, together with changing attitudes towards the role of government, began to merge and create the basis for a change in governance. As the 1992 Industry Commission noted in its inquiry into water resources and waste water disposal:

> Many of the lessons that have emerged from the Commission's previous reports into government business enterprises – such as the need for a more commercial focus in these enterprises – are clearly relevant to the water sector. Indeed, the impetus for many of the reforms now occurring in the sector is a recognition by governments that past policies, which have isolated consumers and service

[150] Anderson, 'If Hayek and Coase were environmentalists'; Ostrom, 'The challenge of common-pool resources'.

[151] Borland, 'Microeconomic Reform', 419–20.

providers from commercial realities and channelled taxpayers' funds into investments with little prospect of providing a worthwhile return to the community, have served Australia badly.[152]

These changes became still more systematised as part of a national competition policy in the early 1990s, led by the Federal Labor government. The national competition policy, the Industry Commission's 1992 inquiry into water services, and the intergovernmental Council of Australian Governments' 1993 review of water resource management recommended the use of market-based processes and together accelerated the shift from government to private sector provision of water services. Behind these polices was the argument that water was a critical resource for the Australian economy. Unless water was provided as efficiently and effectively as possible, national competitiveness, especially in agriculture but also in the manufacturing industry, was at risk, proponents of the reforms argued. In practical terms, this required full cost recovery of the cost of managing and delivering water: a minimum of two-part tariffs for urban water (i.e. an access fee and a usage charge) and either transparency in, or complete removal of, cross-subsidies.

At the State level, the process of corporatising departments, commercialising incentives in services, and reducing government intervention proceeded apace.[153] The swing to smaller government saw departments set new objectives. Where previously public service, public safety, social accountability, and political inscrutability had been the overriding concerns, now the targets became increasing efficiency, productivity, and commercial return. Water supply and engineering departments, long dominated by engineers and planners, were to either adopt more corporate values or, in some states, find parts of their operations privatised. The introduction of price competition and explicit rates of return during the 1970s and 1980s also meant reframing citizens as customers and public utilities as private service providers. To avoid the criticism that public monopolies were to become private monopolies, proponents of change argued for market segmentation and the separation of asset control from service provision. Critics highlighted the long-term costs of asset maintenance in situations where there was no longer an income stream from service provision. As the Binnie and McKinsey reports to the Western Australian and New South Wales governments demonstrated, exemplars from France, the US, and the UK were given to showcase the possibilities

[152] IAC, *Industry Commission*, 1.28. [153] Hilmer, *National Competition Policy*.

of reform.[154] By the late 1990s, state-owned water and sewerage networks in each of the five cities provided services through a mix of privately owned and government-controlled corporations.[155]

Outbreaks of cryptosporidium and giardia in Sydney's water supply in 1998 tested the state's recent reorganisation of its water management. From late July to late September, three million of Sydney's four million residents were urged to boil drinking water to protect their health from harmful parasites.[156] Although Sydneysiders were unscathed by the incident, the health concerns had been real enough: in 1993, a cryptosporidium outbreak in Milwaukee in the USA claimed the lives of more than a hundred people and made about 400,000 people ill.[157] In the aftermath, the finger of blame pointed to both Sydney's corporatised Water Board and the Prospect water filtration plant, the first privately owned such plant in Australia and the largest in the world.[158] Sydney's first major water project since the construction of the Warragamba Dam, the Prospect works filters about 85 per cent of the city's drinking water.[159] After its opening in 1996, representatives of the joint venture company Australian Water Services (its builders) had boasted, 'with the public sector setting the standards and the private sector taking all the commercial risks, it seems that the NSW Government has finally found the best of both worlds'.[160] The subsequent McClellan inquiry into the crisis found otherwise. Although the inquiry was unable to reach a conclusion as to the specific source of the contamination, McClellan noted that in its tender for the Prospect plant, Sydney Water had been 'concerned more with the lowest price than ensuring the highest quality technology'.[161]

That the Sydney water crisis followed so closely after Adelaide's 'Big Pong' of 1997 raised eyebrows among critics of the restructuring of

154 Paterson, 'The privatisation issue'.

155 See Water Services Agency of Australia, 'Water – enriching life and liveability', www.wsaa.asn.au/, accessed 5 June 2021.

156 See New South Wales Premier's Department, *Sydney Water Inquiry Final report*, 1998, where the Commissioner noted 'no identifiable illness' but urged the need for public disclosure whenever evidence of compromised water quality emerged. See also Sheil, *Water's Fall*, on the contamination crisis, which he primarily ascribes to corporate cost-cutting.

157 'Minister pleased at our good health', *Sydney Morning Herald*, 7 September 1998.

158 Murray Hogarth, 'The story of eau', *Good Weekend*, 26 September 1998.

159 'Prospect looks bright for Sydney water bills', *Sydney Morning Herald*, 4 November 1996.

160 'The best of both worlds', *Sydney Morning Herald*, 4 November 1996.

161 McClellan, *Sydney Water Inquiry*, vol. 5, 174.

Australia's urban water sector and the privatisation of public services more generally.[162] For three months, the stench of sewage had engulfed Adelaide, as a foul plume descended from the Bolivar waste water treatment plant.[163] An independent investigation found that the cause was not unseasonal weather, but rather a failure of the facility and inadequate monitoring. Just fourteen months had passed since the South Australian government had contracted the management of the city's water and sewerage system to an international private consortium, United Water International, headed by Thames Water and Vivendi. The size of the contract was unprecedented in Australia and reportedly the largest signed anywhere in the world in 1995.[164] Investigators subsequently found that the filtering processes and stabilisation lagoons had been progressively overwhelmed, leading to the breakdown of bacteriological treatment. 'This heavy loading', the investigation concluded, 'has come about not because of plant overloading but largely because of action to reduce biofilter odour production and operating and maintenance costs'.[165] As Ken Hartley, the lead investigator, later noted, 'the Bolivar incident is an illustration of what can happen when things like monitoring and maintenance are cut to the bone'.[166]

The only winners from the Sydney debacle were bottled water companies and the manufacturers of household water filters. As one distributor boasted, 'the [Sydney] water crisis was the best thing that ever happened to bottled water'.[167] It was already one of the fastest growing segments of the retail market, thanks in part to concerns as to the composition of urban drinking water supplies.[168] In late 1987, Melbourne's *Good Weekend* magazine had published an investigation into the nation's water supplies under the headline, 'Australian water: why it's a worry'. Recent record rainfall in Sydney had precipitated the need for residents in the outer western suburbs to boil their drinking water, while a 1985 report had found half of Melbourne's water supplies were below World Health Organization standards. Echoing the concerns that Carson had earlier raised in *Silent Spring*, the article pronounced, 'questions are being asked about pesticide contamination and the possible

[162] For example, Sheil, *Water's Fall*; Spoehr et al., *State of Secrecy*; Quiggin, 'Contracting out'.

[163] Birnbauer, 'The big pong'.

[164] Sheil, 'An incomplete hypothesis'.

[165] Hartley, *Independent Audit*, 57.

[166] Cited in Birnbauer, 'The big pong down under'.

[167] 'Wet wet wet', *Sydney Morning Herald*, 6 July 1999.

[168] 'Private sector set to clean up', *Sydney Morning Herald*, 24 July 1994; Andrew Hornery, 'Broken public trust will take years to repair', *Sydney Morning Herald*, 1 August 1998.

harmful effects caused by chemicals deliberately added to make water "safe to drink"'. These 'possible harmful effects' included cancer, skeletal fluorosis, Alzheimer's disease, and kidney problems.[169]

The newspaper report was another salvo in an ongoing debate as to the public health implications of chlorinating and fluoridating the nation's drinking water supplies, the latter of which had largely enjoyed the support of scientific and policy elites in the Anglo-American world since the 1960s.[170] Utilities in Sydney, Melbourne, Adelaide, and Perth had begun to add fluoride to drinking water from the mid-1960s to improve dental health (Brisbane did not do so until 2008).[171] Chlorine had been added to urban water supplies from at least the early 1950s to kill bacteria harmful to human health. The National Health and Medical Research Council had published its first guidelines for drinking water in Australia's capital cities in 1972.[172] Water managers in Adelaide responded to this emerging national consensus by seeing their role as to attend not only to resource development, use, and conservation but also to water quality and ecological value. Adelaide had retained rainwater tanks much longer than the other capitals, not least because the city's piped water supplies were unfiltered and had an unpalatable taste and appearance.[173] The construction of several water treatment plants in the late 1970s contributed to a shift away from the suburban reliance on backyard rainwater tanks as tap water quality slowly improved.[174] Nevertheless, in 1989, the State's Engineering and Water Supply Department reported that water quality was the 'single biggest factor' affecting South Australia's present and future water supply.[175]

The Australian Consumers' Association reassured urban Australians in 1991 that 'drinking water from the tap is not the major health risk many believe it to be' but conceded that some areas failed to consistently meet the standards of any drinking water guidelines.[176] The quality of tap water remained a subject for concern, not least as the question of adequate hydration, and its association with human health and wellness, had emerged by the late 1980s.[177] Although medical evidence for such associations was unclear, health columnists enthusiastically encouraged

[169] 'Australian Water: why it's a worry', *Age Good Weekend*, 20 November 1987.
[170] Zelko, 'Optimizing nature'. [171] NHMRC, *Water Fluoridation.*
[172] NHMRC, *Australian Drinking.* [173] Gunn et al., *Benchmarking*, 12.
[174] Engineering and Water Supply Department, *Water Resources*, 20.
[175] South Australia Engineering and Water Supply Department, *Water South Australia*, 3.
[176] 'Australian tap water gets the thumbs-up', *Canberra Times*, 13 August 1991.
[177] See for example Race, 'Frequent sipping', 72–98.

readers to drink at least eight glasses of water a day for a myriad of health benefits, such as weight loss and kidney function.[178] Such advice helped the sales of bottled mineral water to soar from the mid-1970s, when the industry reported a 50 per cent increase in sales in just three years to 1979.[179] Tastes (and trends) mattered too, as discerning palates sought '"a fresher and less chemical taste" than unfiltered tap water', as one Sydney resident observed in 1998 after using a household water filter.[180]

It would take some time for Sydney Water to win back the trust of their customers after the 1998 scare. The proportion of households buying bottled water rose from less than 5 per cent to over 20 per cent during the ten years to 2004, even though newspapers regularly published reports of the more mundane origins of some Australian products.[181]

THE END OF AN ERA

By the late 1960s, Australia had enjoyed a decade of remarkable population and economic growth. While raising living standards and increasing the population, this growth was accompanied by increasing environmental damage and lagging infrastructure provision. Enlivened by their newfound awareness of ecology and a growing movement for Indigenous rights, citizens across the suburban nation joined together to campaign for the protection of urban waterways and culturally significant areas. Although earlier generations had accepted water restrictions in times of drought and outer suburbs that were not connected to sewerage systems, general confidence in engineering solutions, bureaucratic regulations, and governments' policies began to weaken under the environmental, political, and economic challenges then emerging.

Following international trends, alternative methods of resource allocation merged – approaches that used free markets to signal costs, shifted control from public to private providers, and used price to reduce demand

[178] For example, 'Your guide to summer skin', *Australian Women's Weekly*, 10 November 1982, 88; Robin Ingram, 'Water bored: a message in the bottled battle', *Sydney Morning Herald*, 20 September 1988; Pamela Jupp, 'Time to wear less and drink more', *Canberra Times*, 29 December 1991. On the unclear scientific origins of this advice, see Valtin and Gorman, 'Drink at least'.

[179] 'Mineral Water', *Australian Women's Weekly*, 12 September 1979, 20.

[180] David Meagher, 'Getting fresh', *Sydney Morning Herald*, 10 September 1998.

[181] Steen, 'Bottle vs tap', *Choice*. See for example 'Mineral waters from "factory tap"', *Sydney Morning Herald*, 8 August 1982; Alicia Larriera, 'Trendy water springs from household tap', *Sydney Morning Herald*, 14 February 1996.

and raise revenue. The shift to a neoliberal frame of thinking changed not only how water, sewage, pollution mitigation, and infrastructure construction were supplied on a daily basis but also how familiar problems of inadequate supply, poor water quality, pollution, or inadequate infrastructure were conceptualised and answered. The decades from the 1970s to the 1990s saw a wrench in Australians' post-war expectations of how water was supplied and wastewater removed. The shift changed not only who oversaw the provision of water and sewerage services but also the reasons motivating their supply. The relationship between citizen and government was replaced with a relationship between customer and either a public or private corporation, albeit with electoral implications. The price for clean water, sanitary sewerage services, and more limited environmental impacts were passed more immediately and directly to the individual consumer.

Associated with these changes was a long-run decline in per capita consumption of water in each of the five cities. The implication was that in at least one respect, the introduction of clearer price signals had reduced individual demand for potable water supplies and sewage. Other issues, such as how equitable the impact of these changes was on the community, have been less keenly observed.[182] The onset of a nationwide drought in the mid-1990s and the spectre of climate change would soon test these new approaches to Australian urban water management.

[182] For one study see Satur, 'Social inequality'.

8

Twenty-First Century Australian Cities

After two weeks of competition, the world tuned in to watch the closing ceremony of the 2000 Sydney Olympic Games. For Australians, it had been a fortnight of national pride on the world stage – Indigenous sprinter Cathy Freeman had won the women's 400 m final after lighting the Olympic torch, while a teenage swimmer, Ian Thorpe, was the most successful athlete at the event. At Stadium Australia on Homebush Bay, athletes celebrated amidst a spectacular tribute to the nation's suburban kitsch, with racing lawnmowers and pyrotechnic clotheslines.

This finale was not just a farewell to the Games; it was a homage to the post-war vision of the Australian suburbs. During the decade that followed, the nation's lawnmowers fell silent as the Millennium Drought (1996–2010) took its toll on what was left of grassy yards. Australian houses were growing steadily in size, while lots were shrinking as suburbs became more compact. 'Something new has happened to the structure of our cities over the last two decades or so', urban planning critic Bill Randolph observed in 2004, 'that can be seen to represent a threshold between earlier phases of urbanisation and what we might, for want of a better term, call 21st century Australian cities'.[1] Homebush Bay exemplified these developments: once home to the state-owned brick-works and abattoir, the Olympics bid had accelerated the area's redevelopment during the 1980s and 1990s. Sydney's had also been the 'Green' Games, with 450 hectares of parkland set aside to protect wetlands and endangered woodlands, as well as a recycled water system for the

[1] Randolph, 'The changing', 482–83.

FIGURE 8.1. 2000 Olympic Games brochures (Museum of Applied Arts and Sciences).

Olympic Village, which became a model of residential development for other deindustrialising Australian cities to emulate (Fig. 8.1).[2]

In a 2006 report, 'Australia's climate is changing Australia', the Wentworth Group of Concerned Scientists alluded to these urban design approaches. Formed in 2002, just as an El Niño event took hold over the already parched south-east of the continent, the independent body of researchers addressed the environmental challenges of water use at a time of diminishing supply. 'Australia is running out of water', the Group warned, underscoring the urgency of the message with a series of graphs that depicted the steady decline of urban water storages since the mid-1990s.[3] To ameliorate the crisis, the Group urged Commonwealth and State governments to adopt the market reforms, pricing mechanisms, and regulatory tools in the 2004 intergovernmental National Water Initiative.

The approach was influenced by economic rationalist thought on urban water management that was shifting more of the responsibility for water conservation onto residential users, now cast as consumers. New technology and infrastructure promised drought-proofed cities. By the time Freeman lit the Olympic torch, a record drought was being felt

[2] Searle, 'The long-term'. [3] Wentworth Group, 'Australia's climate', 9.

nation-wide. Just as the severe droughts unfolding at the time of the centenary of British colonisation in 1888, the nation's federation in 1901, and during World War II became known as the 'Centennial', 'Federation', and 'World War II' droughts, the drought at the turn of the twenty-first century earned the moniker of the 'Millennium Drought'. Having developed in the mid-1990s, the Millennium Drought finally broke in 2010, at least in south-eastern Australia, as deluge replaced dry and dust. It was the most severe drought experienced in southern Australia since records began in the early twentieth century, thanks to a combination of natural variability and anthropogenic climate change.[4]

The urgency of measures that had characterised the drought years subsequently gave way to more pressing matters of the electoral cycle. Drought would return and will continue to do so; a warming climate contributes to a drying trend across the southern capitals and is likely to exacerbate the impact of ENSO.[5] The long drought left its mark on planning, as impulses of sustainability and efficiency manifest in efforts to shape the urban form as a water catchment of its own. As Australian cities continue to grow, it remains be seen as to whether these plans materialise beyond model suburbs and local initiatives and what their implementation might mean for the water infrastructure and cultures of the past.

DROUGHT IN THE SUBURBS

The dry winter of 2001 delivered the lowest stream flows to Perth's dams since the drought of 1914. Although the state government had promised to 'drought-proof' the city the previous year, its Water Corporation now had to increase its draw on groundwater reserves and tighten water restrictions. The use of sprinklers was prohibited during the day, limited to twice a week.[6] Owners of private bores were also expected to tighten their belts for the first time, with their water use restricted to night-time. The government also targeted Perth's thirstiest residents with a price hike for those households who used more than 550 cubic metres of scheme water a year.[7] Tighter restrictions and higher prices reduced water

[4] Cai et al., 'Did climate', 1–19. [5] McPhaden, Santoso, and Cai, 'Introduction'.

[6] Carmelo Amalfi, 'Water restrictions to hit bore owners', *West Australian*, 4 September 2001.

[7] Steve Pennells, 'Charges aim at top water guzzlers', *West Australian*, 11 February 2003.

consumption in Perth from over 180 cubic metres per capita in 2000–01 to 150 cubic metres the following year.[8]

As the drought gripped other parts of the continent, Sydney and Adelaide implemented water restrictions in 2003, as did Melbourne and Brisbane by late 2006.[9] Compared to these cities, Perth escaped lightly. Sprinklers were allowed once a week in Adelaide and not at all in Sydney, Melbourne, and Brisbane. Washing cars, spraying driveways, and filling swimming pools was out of the question, while Melbourne banned the watering of lawns entirely.[10] Water inspectors kept watch over Melbourne suburbs to ensure compliance, while the New South Wales government encouraged Sydney households to report ('dob in') neighbours who flouted the policies.[11]

Private water sources offered a measure of independence. In Perth, the backyard bore was the preferred option; more than a quarter of households reported in 2010 that they had one, compared to fewer than 10 per cent investing in a rainwater tank.[12] Bores could run year-round, while rainwater tanks were weather-dependent. Seasonal conditions were more conducive to tanks in Brisbane and Sydney, where residents and businesses took up State government incentives to install them. Householders generally understood these 'independent' supplies as strictly for outdoor purposes, maintaining the position that, like grey water, such untreated water was 'dirty'.[13] Water utilities had assumed that the installation of rainwater tanks would facilitate a reduction in water use, but some surveys suggested otherwise. Households with and without tanks in the Illawarra region (a coastal area 100 km south of Sydney supplied by Sydney Water) demonstrated a similar reduction of water use.[14] Bore users in Perth understood their use of independent supplies in similar ways, an approach reinforced by the state government since the late 1990s.[15] As a local resident observed in 2015, 'we don't actually water the garden that much and just use the bore'.[16]

[8] Australian Academy, *Perth's Water Balance*, II. [9] Chong et al., *Review*.

[10] 'Melbourne set for stage 3 water restrictions', *ABC News*, 17 December 2006, www.abc.net.au/news/2006-12-17/melbourne-set-for-stage-3-water-restrictions/2155686, accessed 7 April 2021; 'Toughest water restrictions even for Sydney', *Sydney Morning Herald*, 31 May 2005; Tony Moore, 'Level six restrictions here to stay', *Brisbane Times*, 11 January 2008; Chong et al., *Review*.

[11] 'Vic water wasters could see supply cut', *Sydney Morning Herald*, 29 December 2006.

[12] ABS, 'Water choices'. [13] Moy, 'Rainwater'; Gardiner, 'Do rainwater tanks'.

[14] Moy, 'Rainwater'; Head et al., 'Zones of friction'. [15] Morgan, *Running Out?*

[16] Cited by Lindsay and Supski, 'Changing household water', 54.

As noted in Chapter 6, water restrictions were a tried-and-tested strategy to alleviate pressure on Australia's urban water supplies. What was new this time was the concerted effort to curb water use inside homes and businesses. State governments and utilities offered modest rebates (and plumbing services) to encourage the installation of less thirsty showerheads, washing machines, and toilet cisterns.[17] For guidance, prospective consumers could turn to the Water Efficiency Labelling Scheme, a world-first national scheme introduced in 2004 that required mandatory water efficiency labels on these kinds of fittings and appliances.[18] Of these, it was the dual-flush toilet that was particularly popular in Australian suburbs. Pioneered by Adelaide-based company Caroma in 1980, the dual-flush toilet allowed the user the choice between a button for a full flush (initially, 11 litres) and another for a reduced flush (5.5 litres).[19] By 2007, over 80 per cent of Australian households had at least one dual-flush toilet.[20]

The introduction of water restrictions has historically rendered 'visible' (albeit briefly) the natural and social processes that form the Australian home as a 'project for privacy', cocooned from the outside world.[21] Inside the middle-class suburban home, the everyday experience of water was largely uninterrupted, with the illusion of endless supplies maintained by the household's plumbing and appliances. For most urban Australians, water's easy accessibility indoors facilitated the post–World War II rise in domestic water consumption, as bodies and clothing could now be washed regularly – often more than once a day.[22]

Efforts to curb indoor water use also reflected the changing face of the suburbs. Larger houses and smaller lots were a product of changing lifestyles, the property market, and urban planning mandates. In all Australian cities, the average size of new houses had increased by nearly 50 per cent, from 160 square metres in 1984–85 to 240 square metres in 2006–07.[23] Many critics decried the rise of the latest culprit of Americanisation, the so-called McMansion, and the associated demise

[17] See for example Turner et al., 'Have we achieved'. [18] ABS, 'Australia's environment'.
[19] State Library South Australia, 'Innovations; dual flush toilet', *SA Memory*, 2006, www.samemory.sa.gov.au/site/page.cfm?u=318, accessed 7 April 2021.
[20] ABS, 'Water conservation'.
[21] Kaika, 'Interrogating the geographies', 266; Peel, 'Between the houses', 277.
[22] Davison, 'Down the gurgler; Stanes, Klocker, and Gibson, 'Young adult households.
[23] Dowling and Power, 'Sizing home', 605–19.

of Walter Bunning's 1945 vision of *Homes in the Sun*.[24] Other Australians were opting for a seemingly more cosmopolitan inner-city life – by 2016, there was an apartment for every five houses in Australia, compared to one in seven in 1991.[25]

Regardless of housing type, per capita water consumption was similar inside the home and more closely correlated with socio-economic status and household size.[26] Some utilities sought to disrupt everyday water use by providing residents with shower timers.[27] Focusing on the four-minute shower was at the heart of the Queensland Water Commission's *Target 140* campaign, which aimed to reduce daily per capita water consumption over eight months from 180 litres to 140 litres by the end of 2007.[28] Bright blue timers were distributed to over 1 million Brisbane households, where the Commission's research had found showers were typically seven minutes long and represented over a third of household consumption.[29]

Businesses were also expected to curb water consumption. In Sydney, where industrial and commercial users accounted for a third of the city's water supplies, the local utility sought to improve their water efficiency to help meet the requirements of the utility's operating licence, which mandated 30 per cent reduction of water drawn from all sources between 1991 and 2005 and a 35 per cent reduction by 2011.[30] In Sydney and Melbourne, utilities engaged with firms in planning to reduce water use, with the use of water for production purposes exempted from conservation measures. In the hospitality sector, some business owners resented the additional burden of water conservation, which they viewed as an extension of 'government bureaucracies' and 'more laws'.[31] For all the criticism of these rationing measures, the ends appeared to have justified the means, at least in the short term.[32] Average household water consumption per capita nationwide declined by 14 per cent between 2000 and 2009 and by a further 16 per cent by 2009.[33]

[24] For critique of the McMansion, see Allon, *Renovation Nation*, 141. On shrinking garden space, see Hall, *The Life*.
[25] Davison, *City Dreamers*, 102; ABS, 'Apartment living'.
[26] Troy, Holloway, and Randolph, *Water Use*.
[27] See for example Spearritt, 'The water crisis', 20. In 2007–08, BCC distributed more than 4500 shower heads; BCC Annual Report 2007–2009, 29.
[28] Walton and Hume, 'Creating positive', 215–24.
[29] Walton and Hume, 'Creating positive', 215–24.
[30] Sydney Water, 'Sydney Water's Submission'.
[31] Alonso, 'How Australian hospitality'. [32] Crase and Cooper, 'The political economy'.
[33] ABS, 'Drought drives down'; ABS, 'Water consumption down'.

DROUGHT-PROOFING THE SUBURBS

In an address to Canberra's National Press Club in early 2007, the conservative Prime Minister John Howard was critical of State governments, which had failed to 'significantly augment their water supplies for several decades'. The imposition of water restrictions, in Howard's view, was an indication of the 'foolhardy' nature of this approach. 'Permanent water restrictions should be no more acceptable in our great cities than electricity rationing.' Although the Prime Minister's speech focused mostly on his government's plans for the drought-stricken Murray-Darling Basin, he was adamant that 'we have the capacity to drought-proof our cities'.[34]

The Prime Minister's comments echoed questions from both the left and right of the political spectrum as to whether water restrictions might have been avoided. Favouring more climate-sensitive pricing to instead regulate urban water demand, economists argued that water restrictions were proving financially costly – reducing water utility revenue by around A$150 per household in 2005.[35] Water was valued too cheaply, and as a result, its allocation or distribution was inefficient. This assessment appealed to both the Australian Conservation Foundation and the Productivity Commission.[36] More populist critics argued that water restrictions would not be necessary had the States invested in dam infrastructure rather than folding to environmentalist pressures.[37] In their defence, State governments pointed to the unrivalled severity of the Millennium Drought and to the utility of water restrictions – by reducing demand, they would slow the depletion of dams and allow them to replenish, while raising awareness about water conservation.[38]

The only city spared the Prime Minister's consternation was Perth. Between 2001 and 2006, the Western Australian government had invested twice as much (per capita) in the development of water supplies

[34] John Howard, *A National Plan for Water Security: Transcript of Address to the National Press Club*, 25 January 2007, https://parlinfo.aph.gov.au/parlInfo/search/display/display.w3p;query=Id%3A%22media%2Fpressrel%2FK81M6%22;src1=sm1, accessed 7 April 2021.

[35] See for example Brennan, Tapsuwan, and Ingram, 'The welfare costs'; Grafton and Ward, 'Prices versus rationing'.

[36] Australia Productivity Commission, *Towards Urban*.

[37] Andrew Bolt, 'Not a drop to drink', *News.com.au*, 2 September 2009, www.news.com.au/news/not-a-drop-to-drink/news-story/3bc53fa5f3bf14ae3981b719eb675318?sv=c32897b93f221d56e9b468079209f553, accessed 7 April 2021.

[38] Crase and Cooper, 'The political economy'.

as its counterparts in the four other cities.[39] Increasingly dry conditions had fixed the attentions of the Labor government, since its election in 2001, on Perth's emerging water challenges. As the drought depleted the city's dams and groundwater supplies, the Gallop government announced in early 2002 its intentions to invest in seawater desalination technology to supply water for the suburbs. A reverse osmosis desalination plant would be constructed in the city's industrial heartland to provide potable water during periods of extreme drought and avert total sprinkler bans.[40] No Western Australian government wanted to repeat the bans of the 1970s.[41] The reverse osmosis desalination of seawater had been on the state's long-term agenda since the late 1980s, when the then Water Authority had concluded that it was preferable to obtaining water from farther afield.[42] Then it had been too expensive to warrant investment, especially when more cheaply accessible groundwater supplies would suffice, at least in the short term.

Lacking support from environmentalists and the Coalition opposition, the government initially explored alternatives, such as extracting water from rural aquifers and transferring water from the state's north-west. But as nationwide anxieties about drought and climate change grew, proponents of desalination held the trump card: seawater desalination would offer Perth a 'climate-independent' water supply.[43] Australian climate researchers had recently concluded that Perth and the surrounding southwest of the continent had undergone a drying trend since at least the mid-1970s, which they attributed to anthropogenic climate change.[44] Tim Flannery, a mammologist, environmentalist, and writer on climate change, predicted that 'Perth will become a ghost metropolis over the next few decades unless governments acknowledge that global warming is a reality'.[45] The prevailing dry conditions (and associated restrictions on household water use) helped to substantiate this prediction, which fed concern about the prospect of scarce water resources in the future.

Desalination offered a solution to protect Perth from such vagaries of the weather, as the director of Western Australia's water strategy proclaimed in 2004: 'We need a water supply which is independent of

[39] Marsden and Pickering, *Securing Australia's*, ES.ii.

[40] Environmental Protection Authority, 'Perth Metropolitan'.

[41] See Morgan, *Running Out?* [42] Mauger, *Planning Future*, 14.

[43] Wendy Pryer, 'Garrett lends hand as Gallop changes tack on Greenhouse', *West Australian*, 15 September 2004.

[44] Indian Ocean Climate Initiative (IOCI), Climate Variability; Morgan, *Running Out?*

[45] Carmelo Amalfi, 'Perth will die, says top scientist', *West Australian*, 25 June 2004.

climate.'[46] Two years later, Perth residents became the first in Australia to drink desalinated seawater, and a second desalination plant was commissioned the following year.[47] Other state governments took note, not least as Flannery warned in an oft-misquoted television interview: 'We're already seeing the initial impacts [of climate change] and they include a decline in the winter rainfall zone across southern Australia ... even the rain that falls isn't actually going to fill our dams and river systems.'[48] For their part, the Wentworth Group recommended that 'desalination, potable reuse, stormwater capture, all need to be recognised as viable economic options to be costed properly. Often they are better options than building more dams and damaging more coastal rivers.'[49] Before long, desalination plants were on the drawing board for Sydney, Melbourne, Adelaide, and Brisbane.[50] No longer dismissed as 'bottled electricity', as former New South Wales Labor Premier Bob Carr had once quipped in reference to its financial cost, desalination represented to governments seemingly limitless water and, perhaps most importantly, electoral security.[51]

Despite the widespread appeal of desalination to state governments and their water utilities, the economic and environmental implications of these decisions did not go unquestioned. A 2005 poll commissioned by SCUD (Sydney Community United against Desalination) found 60 per cent of Sydney residents were opposed to the decision to build a desalination plant.[52] A subsequent parliamentary inquiry concluded it would be unnecessary to build such a plant if the government instead pursued water recycling and reuse strategies.[53] The inquiry's chair questioned the motives behind Sydney Water's plan: 'It actually makes money out of selling water and therefore it's not encouraged ... to act on its water

[46] Carmelo Amalfi, 'Desalination plant', *West Australian*, 28 April 2004.

[47] Amanda Banks, 'Tap turned on at Perth's $387 million desalination plant', *West Australian*, 20 November 2006.

[48] *ABC Landline*, 11 February 2007, www.abc.net.au/local/archives/landline/content/2006/s1844398.htm, accessed 7 April 2021.

[49] Wentworth Group, 'Australia's Climate', 8.

[50] Spearritt, 'The water crisis', 30.

[51] 'To quench our thirst it will cost 255,500 tonnes of greenhouse gases', *Sydney Morning Herald*, 1 November 2004.

[52] Wendy Frew, 'City desalination plant is not the solution – poll', *Sydney Morning Herald*, 14 December 2005; Murray Trembath, 'Fierce opposition to Kurnell desalination plant after 2005 announcement', *St George and Sutherland Shire Leader*, 15 November 2019.

[53] New South Wales Parliament, 'A sustainable water supply for Sydney', General Purpose Standing Committee No. 5 (Sydney: The Committee, 2006).

saving initiatives.'[54] The Victorian government similarly faced a backlash when it announced in 2007 its plan to build a plant near Wonthaggi on the Bass coast. Noting that 'views are divided within the community', a parliamentary inquiry feared that the plant might provide a disincentive for water conservation and the development of other water supplies.[55]

Despite these misgivings, State governments defiantly kept desalination on the table. Some environmental concerns could be allayed by siting the plants in already degraded industrial areas, such as Perth's Kwinana, Adelaide's Port Stanvac, and Sydney's Kurnell. In south-east Queensland and Victoria, protesters fought to protect scenic beaches from the pollution associated with the desalination process, which expels a salty brine into the ocean. In both instances, governments limited their consultation with local communities, citing the pressing need to expedite construction.[56] The Queensland government bypassed its processes of environmental impact assessment to start construction at Tugun in 2007. As bullish Labor Premier Peter Beattie told one local protester, 'if we don't have desal, we're not going to have any water. If you don't have water, you're dead.'[57]

Beyond local environmental impacts, the carbon emissions of desalination plants came under scrutiny. The nation's climate politics had been heating up as the drought wore on, and by 2009, Labor Prime Minister Kevin Rudd could declare that climate change 'is one of the greatest scientific, economic, and moral challenges of our time'.[58] Some dissenters in Perth baulked at a solution to the city's water challenges that would increase the carbon emissions already associated with the regional drying trend.[59] One analyst calculated that the proposed Sydney plant would be the equivalent of adding 220,000 cars to the roads each year in terms of carbon emissions, noting 'any talk of "carbon offsets" is simply a

[54] 'Sydney desalination plant unnecessary: inquiry', *ABC News*, 9 June 2006, www.abc.net.au/news/2006-06-09/sydney-desalination-plant-unnecessary-inquiry/1773814, accessed 7 April 2021.

[55] Environment and Natural Resources Committee, *Inquiry into Melbourne's Future Water Supply* (Melbourne: Government Printer, 2000), 9.

[56] For example, 'Protesters march against desal plant', *ABC News*, 4 September 2008, www.abc.net.au/news/2008-09-04/protesters-march-against-desal-plant/499202, accessed 7 April 2021.

[57] *Courier-Mail* (Brisbane), 1 February 2007.

[58] Kevin Rudd, 'Prime Minister's prizes for science', *PM Transcripts*, 28 October 2009, https://pmtranscripts.pmc.gov.au/release/transcript-16882, accessed 7 April 2021.

[59] Morgan, *Running Out?*

smokescreen to hide the plant's true environmental impact'.[60] Writing in the thick of these debates, one scholar was aghast at Queensland's turn to desalination: 'In going down the high-capital, high-energy and high-carbon emissions road, Queensland has now committed a generation to paying through the nose for a desal quick-fix rather than confronting more sustainable approaches to climate change.'[61]

For New South Wales, Queensland, and Victoria, the drought-breaking rains of 2010–11 put something of a dampener on enthusiasm for desalination plants.[62] Melbourne's plant was the largest and had yet to even commence operations before it was mothballed on its completion in 2012, along with the controversial Sugarloaf Pipeline, idle since 2010.[63] In Sydney, a cash-strapped State government welcomed the long-term lease of its desalination plant (completed in 2010) to a superannuation funds group, even though the deal locked in 50 years of annual fees to be borne by consumers, regardless of whether the plant was operating or not.[64] With water storages near full again, commentators were especially critical of these so-called white elephants, which became political millstones for the governments responsible for their construction.[65] An economic review of the urban water sector in 2011 concluded that many such projects could have been deferred or replaced with alternative, lower cost sources of water.[66]

For Brisbane, the nation's fastest growing city between 1995 and 2006, desalination soon proved to be a lame duck.[67] During the drought, water

[60] Australia Institute, 'Greenhouse implications of the purposed Sydney desalination plant', July 2005, https://australiainstitute.org.au/wp-content/uploads/2020/12/WP78_8.pdf, accessed 7 April 2021.

[61] Spearritt, 'The water crisis', 34.

[62] Cooley and Ajami, 'Key issues', 103.

[63] Adam Carey, 'Government won't turn on $750m pipeline despite dry spring forecast', *Age* (Melbourne), 14 August 2018.

[64] 'Desal deal to pour $2.3bn into NSW coffers', *ABC News*, 11 May 2012, www.abc.net.au/news/2012-05-11/desal-deal-to-pour-242.3bn-into-nsw-coffers/4004918?source=rss, accessed 7 April 2021; Ursula Malone, 'New South Wales desalination plant deal to cost consumers $10 billion over 50 years', *ABC News*, 27 September 2013, www.abc.net.au/news/2013-09-27/nsw-desalination-plant-deal-costing-customers-$10-billion/4985168#:~:text=The%20sale%20of%20the%20plant,more%20than%20a%20year%20ago, accessed 7 April 2021.

[65] See for example Alison Caldwell, 'Victorians stuck with desalination white elephant', *ABC World Today*, 1 March 2011, www.abc.net.au/worldtoday/content/2011/s3151800.htm, accessed 7 April 2021; Andrew Clennell, '$700k a day bill for desalination plant', *Daily Telegraph* (Sydney), 17 May 2011.

[66] Productivity Commission, *Australia's Urban Water Sector: Productivity Commission Inquiry Report* (Melbourne: Productivity Commission, 2011).

[67] See Spearritt and Cook, 'Water forever'.

restrictions, rebates, and publicity campaigns combined to limit per capita water use to a mere 129 litres per day, lower than any other capital. Given the average daily consumption had been over 300 litres per person prior to the drought, this was an impressive result.[68] As water levels in Wivenhoe Dam continued to decline – reaching 15 per cent of capacity in mid-2007 – the Beattie Labor government announced the construction of the Southeast Queensland Water Grid or SEQ Water Grid. This project promised the city 'safe, secure and sustainable water' by connecting all major dams in south-east Queensland, including Wivenhoe. The 600 km network of pipelines soon connected to the 'climate resilient water sources' of the Western Corridor Recycled Water Scheme (completed in 2008) and the desalination plant at Tugun (completed in 2009).[69]

By late 2010, Australia had entered the strongest La Niña phase since 1974, which precipitated record rainfall across the continent's south-east.[70] In the New Year, heavy rains pushed aside fears of scarcity, as a deluge descended on Queensland affected three-quarters of the state.[71] Worse was to come. Within 48 hours between 10 and 12 January, 1000 mm fell in the Brisbane River catchment, and floodwaters soon raced towards the capital. A *Courier-Mail* journalist watched the roaring, reeking floodwaters:

> Its cargo – broken window frames, tyres, plastic bottles, even a fleet of half a dozen navy couch cushions – was, I understood, the sad detritus of people's lives. Our own people's lives, after yet another catastrophic flooding on the Brisbane River.[72]

At their highest, flood waters reached 4.46 m at the port office gauge in Brisbane, flooding over a thousand homes, inundating nearly two thousand businesses across ninety-four suburbs, and taking sixteen human lives (Fig. 8.2).

As the floodwaters receded and recovery got underway, so began a blame game that culminated with a billion-dollar class action lawsuit. Even the Brisbane River itself was seen as culpable, with some locals expressing the feeling that the river had betrayed them. The editor of the *Courier-Mail* wrote: 'That big, brown powerful river. Every time I look at the Brisbane River I will remember what she did to us; what

68 Queensland Water Commission, *The 2008 Water Report*, 6, www.parliament.qld.gov.au/Documents/TableOffice/TabledPapers/2009/5309T449.pdf, accessed 19 April 2021.

69 SEQ Water Grid Manager Annual Report 2009–10, 8.

70 Australian Bureau of Meteorology, *Record-Breaking*.

71 Cook, 'Vacating the floodplain'.

72 Cited in Cook, *A River with a City Problem*, 158.

FIGURE 8.2. Floods in Brisbane, 2011 (photographer: John Doody).

she is capable of.'[73] Another journalist at the *Sunday Mail* suggested that 'it would be only human to feel a sense of betrayal or contempt after the way our river, our beloved river, rose up and struck out so violently at the very people who took it into their hearts'.[74] Many residents had long believed the Somerset and Wivenhoe dams had 'flood-proofed' their city; the latter's construction had promised to 'tame the river' and assured residents that the 'flood threat' was 'past'.[75]

The dams had reduced the 2011 flood's potential height by two metres, but decades of development on the floodplain of the Brisbane River had put residents and businesses directly in harm's way. Successive State and municipal governments favoured urban growth and consolidation over more cautious approaches to flood mitigation by regulating land use.[76] Blaming dam mismanagement for the 2011 disaster deflected attention from poor land-use practices, government fault, and the river's hydrology. As engineer Trevor Grigg noted, the strategy served to reinforce the view that 'if only we get the operation of the dam right, we won't have any flooding of this river'.[77]

[73] Cited in Cook, 'Shaken', 88. [74] Cited in Cook, 'Shaken', 88.
[75] Cited in Cook, *A River with a City Problem*, 128–29.
[76] Cook, 'Vacating the floodplain'. [77] Cited in Cook, 'Shaken', 96.

This engineering focus prevailed with the launch of a class action in 2017. The State's Commission of Inquiry found that while 'the disastrous floods … were unprecedented, in many places completely unexpected, and struck at so many points at once that no government could be expected to have the capacity to respond seamlessly and immediately everywhere', there had been 'non-compliance with the manual under which the [Wivenhoe] dam was to be operated'.[78] This finding became the basis for a class action on behalf of nearly 7000 residents and business owners, which their lawyers pursued against the Queensland government and dam operators in the New South Wales Supreme Court.[79] In early 2021, the Queensland government agreed to compensate the claimants, while the dam operator is appealing against the Supreme Court's decision that its engineers had acted negligently before and during the floods.[80]

Following a familiar pattern of reacting to water crises wet or dry, the urgency that had characterised urban water management for nearly a decade soon waned once the drought broke. By 2014, the National Water Commission, an independent body reporting to the federal government, observed, 'at present water is not a priority for most governments across Australia, partly because of the substantial gains made through the [microeconomic] reform process and partly because the major urban centres and irrigation districts are no longer in drought'.[81] Although these urban water management strategies had forestalled drought, they were soon found to have been costly in other ways. In Queensland, the Newman Liberal–National government baulked at the shortfall in water revenue in 2013 as a result of their predecessors' campaign to curb water consumption. Construction on a water treatment plant was halted, the Water Grid mothballed, and water prices increased to help alleviate the state's budgetary pressures. Households had to foot to the bill with higher water prices, while the government put on hold the construction of

78 Queensland Floods Commission of Inquiry, *Final Report*, 30.

79 The case could progress through the New South Wales courts because a claimant lived there and lost a business to the floods. The Supreme Court of Queensland did not hear class actions when the case was filed.

80 Rachel Riga, 'Brisbane 2011 flood victims win $440 million in class action partial settlement over operation of Wivenhoe Dam', *ABC News*, 26 February 2021, www.abc.net.au/news/2021-02-26/brisbane-floods-wivenhoe-dam-class-action-win-payout-queensland/13196490#:~:text=Victims%20of%20the%202011%20Brisbane,%2C%20north%2Dwest%20of%20Brisbane.&text=The%20historic%20settlement%20covers%2050,%24880%20million%20liability%20for%20damages, accessed 7 April 2021.

81 National Water Commission, *Australia's Water Blueprint* (Canberra: National Water Commission, 2014), 3.

further water infrastructure.[82] Local poet and novelist Thomas Shapcott observed during the 2011 floods: 'Long after the cameras and media have lost interest, the victims will still be cleaning up. … But there will be people who still get strange feelings whenever it rains. … Floods may go down, but the things they leave behind take much longer.'[83]

LIFE ON THE WATER

One of the enduring images of the 2011 Brisbane floods was the city's controversial Riverwalk breaking apart and threatening to collide with the city's Gateway Bridge as it careened down the Brisbane River.[84] Connecting the inner northern riverside suburb of New Farm to the city centre, the 850 m floating walkway of concrete pontoons had been constructed in 2003 as part of the *Building Better Cities Program*, an urban renewal project established by the Commonwealth Hawke Labor government in 1991 to 'promote improvements in the efficiency, equity and sustainability of Australian cities'.[85] A rare intervention of a Federal government in water policy, its funding supported improvements to the ageing or inadequate water infrastructure of de-industrialising suburbs across the nation, such as sewerage and water systems in Sydney's Ultimo-Pyrmont area and Fremantle, flood mitigation works in inner Melbourne and outer Brisbane, and stormwater drainage in Elizabeth, an Adelaide suburb.[86]

Australian cities were undergoing a transformation; in a globalising era, manufacturing industries were subject to job losses due to tariff reductions, outsourcing, and automation since the economic reforms of the 1980s. The impact of deindustrialisation re-shaped central and inner suburban areas, as the growth of the service sector, the rising importance of international higher education, and the development of leisure and

[82] APN Newsdesk, "Qld Water charges going up to pay off Labor's $7b grid", *Sunshine Coast Daily*, 3 May 2013, www.sunshinecoastdaily.com.au/news/qld-water-charges-going-pay-labors-7b-grid/1853892/, accessed 7 April 2021.

[83] Thomas Shapcott, 'Nowhere to hide from the mayhem when the river comes looking for you', *Sydney Morning Herald*, 15 January 2011.

[84] Jessica Hinchcliffe and Terry Bagley, 'Where did the original Brisbane Riverwalk end up after the 2011 floods?', *ABC News*, 21 July 2016, www.abc.net.au/news/2016-07-21/flooded-brisbane-riverwalk-given-second-life/7645566?utm_source=abc_news_web&utm_medium=content_shared&utm_content=link&utm_campaign=abc_news_web, accessed 7 April 2021.

[85] Cited in Auditor-General of Victoria, *Building Better Cities*, 11.

[86] Freestone, *Urban Nation*, 232–33; Bryant, *Investing*, 34–35.

lifestyle precincts prompted a boom in apartment construction and the repurposing of worker cottages, factories, and commercial buildings for middle-class housing. 'Perhaps for the first time ever', observes historian Seamus O'Hanlon, 'the rapid growth of this housing sector has seen the emergence in Australia of a genuine urban sensibility, most noticeably in Sydney and to a lesser extent in Melbourne and elsewhere'.[87]

In these post-industrial cities, proximity to water, either riverine or harbourside, proved especially attractive to urban planners and property developers. No longer sewers for the city, waterscapes such as the Brisbane River, Melbourne's Yarra, and, most famously, Sydney Harbour, have become centrepieces for the nation's growing global cities. This has been no accident. Realising in the early 1980s that the visual appeal of the city was crucial to attracting global capital, State and municipal governments collaborated with investors and developers to makeover the rivers, bays, and docklands of the Australian capitals.[88] A one-kilometre stretch of the Yarra was transformed from a bleak industrial area to 'Southbank' – a cultural, business, retail, and residential hub intended to be appealing and accessible to locals and tourists. By the mid-1990s, this remade waterfront area had become a magnet for apartment building, as the State government embraced the market and the real estate appeal of living on the river – now much cleaner, thanks in part to the 'Give the Yarra a Go' campaign of a local newspaper a decade earlier.[89]

The ongoing appeal to private interests resulted in the social polarisation of these waterfront areas, as urban critic Kim Dovey shows in his assessment of the Yarra redevelopment, which by 2005 extended to the city's 'Docklands'. The 'vast majority of the Melbourne population cannot afford to buy or rent Docklands housing and parking costs make a visit expensive'.[90] The redevelopment of East Perth from neglected suburb to fashionable inner-city address by the late 1990s similarly came at a significant cost. At the new engineered Claisebrook Lake and Claisebrook Cove, there were soon few traces of the low-rent houses, hostels, and services that had sustained Perth's low-income and transient residents since the 1950s.[91] Aboriginal people were especially affected, having lived in the area prior to colonisation and since, while also

87 O'Hanlon, *City Life*, 5–6. 88 Dovey, *Fluid City*, 33.
89 Dovey, *Fluid City*, 33–38, 86–92. 90 Dovey, *Fluid City*, 198.
91 Gregory, 'Obliterating history?'; Gaynor and Martin, 'Every fountain tells a story'.

accessing the specialist services established there, such as public housing, medical care, and education.[92]

In step with other waterfront developments of formerly industrial areas, much of the built heritage of East Perth was erased through the suburb's renewal and gentrification – driven by the need to sell housing to pay for the beautification of the precinct. Contributing to this appeal were artistic uses of water, such as 'The Life of the Claisebrook' sculpture, which typify urban renewal projects across Australian cities and suburbs.[93] Through local applications of the US urban design doctrine New Urbanism since the 1990s, critics argue that these master-planned developments relegate water to an aesthetic and static device – part of an image of a sustainable community that appeals to the market. 'Sustainability' in this New Urban context is equated with walkable and 'transit-oriented' communities, rather than those that integrate or sustain ecological and hydrological processes.[94] Such examples of New Urban water reflect what Kaika and Swyngedouw describe as an alienation from infrastructural and natural processes that have been rendered invisible, acutely expressed in the context of escalating global environmental issues.[95]

'The Life of the Claise Brook' sculpture was produced in consultation with a Noongar elder, while other public artworks in the area were produced by Noongar artists and refer to Noongar stories. Yet, as one community member noted in the context of a later review of the project, the redevelopment project had 'done exceptionally well in acknowledging the significance of the area in pre-European times but much less well in embracing the history of Aborigines in the very recent past'.[96] East Perth is not alone in this regard: literary scholar Emily Potter describes a park named for a Kaurna woman ('Lartelare') in Adelaide's New Port Quays as 'a gesture of remembrance, but a far cry from political and legal enfranchisement, and the historical recognition this entails'.[97] Such public art and Aboriginal place names are part of a wider settler phenomenon in Australian cities, which historian Graeme Davison understands as 'rescuing them from the curse of recency'.[98] Since the 1990s, the proliferation of Aboriginal words for sites such as Birrarung Marr, a riverside park on the

92 Hillyer, 'Bennett House', 147–82.
93 Gaynor and Martin, 'Every fountain tells a story'.
94 Benac, 'The New Orleans', 388–403.
95 Kaika and Swyngedouw, 'Fetishizing', 120–38.
96 East Perth Redevelopment Authority, *Claisebrook Village*, 12.
97 Potter, 'The gathering storm', 231.
98 Davison, *City Dreamers*, 259.

banks of Melbourne's Yarra, and Barangaroo on Sydney's waterfront, reflect an ongoing project of reconciliation between Indigenous and non-Indigenous Australians.

Among the champions of Sydney's Barangaroo waterfront redevelopment is Paul Keating, who was Labor Prime Minister from 1991 to 1996. Keating's 1992 'Redfern speech' was delivered in a suburb 'just a mile or two from the place where the first European settlers landed, [which] in too many ways tells us that their failure to bring much more than devastation and demoralisation to Aboriginal Australia continues to be our failure'.[99] This recognition of the toll of colonialism on the nation's First Peoples reflected the Keating government's broader support for the legal recognition of the continuing rights and interests of Aboriginal peoples to lands and waters (their 'native title').

This campaign for Aboriginal land rights had been underway since the 1970s along Sydney's Georges River, which local Aboriginal people knew to have a wealth of art sites, stone tools, art-filled overhangs, and campsite remains, such as hearths and middens.[100] Protecting this history and heritage became one of the driving forces behind the establishment of the Gandangara Local Aboriginal Land Council in 1983, which successfully claimed in 1988 an escarpment on Guragurang (Mill Creek). Theirs was the largest holding of land of any of the Local Land Councils in the metropolitan area at that time.[101] In 1992, the State's National Parks and Wildlife Service (NPWS) resumed control of the Georges River parklands and, over the following two decades, expanded them to include the southern bank of the river to Guragurang. Although a joint management arrangement for the parklands between the NPWS and the Aboriginal owners is unfolding only slowly, there are growing numbers of Aboriginal people employed in the NPWS and as park liaison officers.[102]

This bushland in Sydney's south-western suburbs has also become significant to Sydney's more recent migrants of south-east Asian and Arabic background, many of whom have emigrated since the 1970s and settled in nearby suburbs such as Cabramatta, Fairfield, Liverpool, Bankstown, and Lakemba. Some of these migrants have fled conflict and persecution, and water provides a way for them to make 'safe places,

[99] Paul Keating, 'The Redfern Address', 10 December 1992, *Australian Screen*, https://aso.gov.au/titles/spoken-word/keating-speech-redfern-address/extras/, accessed 30 March 2021.

[100] Goodall and Cadzow, *Rivers and Resilience*, 232.

[101] Goodall and Cadzow, *Rivers and Resilience*, 244–47, 52–54.

[102] Goodall and Cadzow, *Rivers and Resilience*, 271–72.

in both health and cultural terms', as observed by Heather Goodall and Hazara social scientist and community activist Latifa Hekmat.[103] Although recent migrants enjoy the reliability and convenience of piped water in suburban Australia, some groups, such as Hazara women in Western Sydney, remain distrustful of the government agencies that provide water and ensure they boil water before drinking. For this community, who tend to rent their homes due to uncertain visa conditions, Australian bathroom plumbing can prove culturally and religiously challenging. Water is a necessary element of physical and spiritual purification after defecation for many religious groups across South Asia and the Middle East. Yet, as Hazara women have found, there are often few taps close to toilets or bidets, and many have had to improvise in their rented Sydney properties.

Recognising the importance of the Georges River to these culturally diverse communities, the NPWS commissioned a study of the cultural dynamics of the parkland's visitation in the early 2000s.[104] Through interviews with park visitors and staff from Aboriginal, Vietnamese, Arabic, and Anglo-Celtic backgrounds, researchers learned that they shared among them a high regard for water in the parklands, which they understood and used in different ways. Muslim women and children have participated in stewardship programs along the Georges River, as they have near Melbourne's Merri Creek, while people of Chinese Vietnamese background understood some areas of the river in terms of feng shui to be luckier than others.[105] Fishing also had widespread appeal, as Mahmoud Dehn remembered in 2003: 'My parents used to take us down to the Georges River to fish, to relax, to just get out of the sort of environment we're in where it's mainly like flats and buildings; you know, just to get out of there, get into nature, have some fun ... learn how to fish'.[106] Such activities play an important social role, as a Vietnamese-Australian grandmother explained:

> To see all your children and grandchildren playing together, bonding, enjoying each other's company. I get to talk about old times, to teach and remind my children and grandchildren of their background and cultural origin. It is also fun

103 Goodall and Hekmat, 'Talking to water'.

104 Byrne, Goodall, and Cadzow, *Place-Making*.

105 Goodall, *Waters of Belonging;* Bush, Miles, and Bainbridge, 'Merri Creek', 170–79; Cadzow, Byrne, and Goodall, 'Waterborne'.

106 Interviewed by Stephen Gapps, 2003, cited in Byrne, Goodall, and Cadzow, *Place-Making*, 72.

sometimes when we catch fish but it rarely happens. However, it's a wonderful feeling to see your family having fun, no more war, no more poverty, no worries about work and all.[107]

Since the 1980s, successive governments have steered Sydney's population growth across the floodplain of the Hawkesbury–Nepean to what is now the Western Sydney region. On Darug country, writes Grace Karskens, 'modern suburbs are built over battlefields'.[108] In 2017, Karskens chanced upon a handwritten nineteenth-century list of nearly 180 Darug place names for Dyarubbin, the Hawkesbury–Nepean River, which she has worked with local Darug women and linguists to map onto the area.[109] Reflecting on this collaboration during the March 2021 floods, Darug author Jasmine Seymour saw the power of eel serpent Garangatty at work: 'He is furious and explodes out of the waterhole in a huge display of flood power.'[110] This return of Darug names to Dyarubbin reflects a wider shift in the nation's water management since the Millennium Drought.[111] The intergovernmental National Water Initiative of 2004 gave recognition, albeit modest, to Indigenous access to water resources and provided for the inclusion of Indigenous people in water planning that 'will incorporate [I]ndigenous social, spiritual and customary objectives and strategies for achieving these objectives wherever they can be developed'.[112] Although the implementation of these principles has fallen well short of the aspirations of Aboriginal peoples, particularly in the Murray-Darling Basin, water utilities in the capitals are increasingly consulting with local Aboriginal knowledge-holders. In 2018, the Victorian Labor government appointed the Birrarung Council, with representatives from the Wurundjeri Woi Wurrung Cultural Heritage Aboriginal Corporation, to serve as the 'independent voice of the Yarra River as a living entity' as prescribed its *Yarra River Protection (Wilip-gin Birrarung murron) Act 2017*.[113]

107 Bach, interviewed by Alison Phan, 2002, cited in Byrne, Goodall, and Cadzow, *Place-Making*, 88.

108 Karskens, *The Colony*, 732.

109 Karskens et al., 'Friday essay'.

110 Cited in Julie Power, 'Putting aboriginal names and places back on the map starts by the river', *Sydney Morning Herald*, 26 March 2021.

111 Taylor, Moggridge, and Poelina, 'Australia Indigenous'.

112 Council of Australian Governments, *Intergovernmental Agreement on a National Water Initiative* (Canberra: Council of Australian Governments, 2004), 9.

113 Victoria State Government, 'Birrarung Council', *Waterways and Catchments*, 20 November 2020, www.water.vic.gov.au/waterways-and-catchments/protecting-the-yarra/birrarung-council-the-voice-of-the-yarra, accessed 7 April 2021.

CATCHMENT CITIES

As South Africa's Cape Town teetered on the brink of 'Day Zero' in early 2018, Australia's urban water managers looked back on the Millennium Drought, taking stock of the state of the water supplies in the capitals. In Western Australia, the McGowan Labor government saw an opportunity to publicise the diversification of Perth's water supplies, with the Water Minister noting that 'we face very similar climatic conditions to Cape Town. … We could have been in a similar situation had we not made decisions a decade ago to invest in climate-independent sources of water'.[114] As the continent's southwest measured its driest June on record, Perth's desalination capacity was expanded in 2013, and the local water utility declared, 'rainfall is no longer relied upon as our primary source of water. Our investment in desalination has secured our water supply in dry years'.[115] Due to the regional drying trend, streamflow into the city's dams was declining to such an extent (from a mean annual streamflow of 338 million cubic litres during 1911–74 to 134 million cubic litres during 1975–2017, of which the period 2012–17 had just 47 million cubic litres) that these reservoirs had become storage for potable water supplies from Perth's suburban aquifers and desalination plants.[116] Since then, a further two desalination plants have been mooted for the city.[117]

For the other Australian capitals, desalination did not displace dams as the primary source of water supplies. In Sydney and Brisbane, desalination plants stood largely dormant since their construction, while their use has been more intermittent in Adelaide and Melbourne. Bushfires in water catchments in 2019 prompted the largest purchase of water from Victoria's Wonthaggi plant.[118] These catchments were still recovering from the toll of the 2009 Black Saturday bushfires, as regenerating forests

114 Irena Ceranic, 'Perth dodged a water crisis like Cape Town's Day Zero-but the city's still not out of the woods', *ABC News*, 21 June 2018, www.abc.net.au/news/2018-06-21/how-perth-dodged-its-own-water-crisis-like-day-zero-in-cape-town/9891472, accessed 7 April 2021.

115 Cited in Daniel Mercer, 'Desal Water Props up Dams', *West Australian*, 3 July 2013.

116 CSIRO and Bureau of Meteorology, *State of the Climate 2018* (Melbourne: CSIRO and Bureau of Meteorology, 2018), 9.

117 Daniel Mercer, 'Water Corporation submits plans for $1 billion desalination plant to fix Perth water supply', *West Australian*, 23 April 2019.

118 Adam Carey, 'Ash from fires threatens Melbourne's water as large desal order placed', *Age*, 18 March 2019.

use more water than the mature forests they replace.[119] Adelaide's plant increased production in late 2019 to alleviate the city's reliance on the Murray and to allow drought-stricken farmers upstream in New South Wales and Victoria to draw more from the river.[120]

The allure of 'climate-independent' water supplies was difficult for State governments to resist during the Millennium Drought, as severe water restrictions rankled voters. For the South Australian government, the idea of 'water proofing Adelaide' and ensuring 'water for good' – as its 2004 and 2010 water planning strategies promised – was especially welcome.[121] In addition to investing in seawater desalination at Port Stanvac, the State has focused on enhancing its collection of stormwater for non-drinking purposes, such as irrigation and industrial uses. First mooted in 1989, there was a 'tantalising fit' of unused resource and demand, as a feasibility report noted in 2004, between the city's stormwater run-off to the Gulf St Vincent (annually between 160 and 250 million cubic metres) and urban consumption (about 200 million cubic litres each year).[122]

South Australia has long sought to make the most of its relatively meagre rainfall and to supplement its reliance on the Murray. Often just a trickle by the time it reaches the South Australian border from its headwaters near Mount Kosciuszko, the Murray has been a source of Adelaide's water supplies since 1954.[123] During the Millennium Drought, when rainfall in the Basin was the second lowest only to the dry years of the World War II, flows across the border were less than 10 per cent of the twentieth-century average.[124] Beyond Adelaide, what little freshwater remains flows to the Ramsar-listed Coorong and Lower Lakes wetlands at the Murray Mouth, near Goolwa, from where it enters the sea. The quantity and quality of water that reaches Adelaide and the Murray Mouth are subject to not only the vagaries of the weather but also the ongoing contestation of flows upstream that pre-date the nation's

[119] Feikema, 'Influence of climate'.

[120] Nick Harmsen, 'How will the revival of Adelaide's desalination plant help drought-stricken farmers?' *ABC News*, 7 November 2019, www.abc.net.au/news/2019-11-07/how-will-the-sa-desal-plant-revival-help-australian-farmers/11682044, accessed 7 April 2021.

[121] Water Proofing Adelaide, *Water Proofing Adelaide: A Thirst for Change* (Adelaide: Water Proofing Adelaide, 2004); Office for Water Security, *Water for Good: A Plan to Ensure Our Water Future to 2050* (Adelaide: Office for Water Security, 2009).

[122] Kellogg Brown and Root, *Metropolitan Adelaide*, vi; Rolls, 'Stormwater management'.

[123] ABS, *Water Supply*.

[124] Kingsford et al., 'Engineering a crisis'.

federation in 1901.[125] Under 'natural' conditions, ecologists estimate, the Murray would likely flow to the Mouth every day; at the end of the Millennium Drought, the river's flow fell short 40 per cent of the time, exposing hazardous acid sulfate soils.[126]

In the 1950s, South Australian State geologists had explored the possibility of recharging the confined aquifer beneath Adelaide's western suburbs with stormwater for later use.[127] But it was flood mitigation and drainage that were the city's immediate concerns. The urban footprint was spreading onto the Northern Adelaide Plains, where farming was giving way to housing a growing post-war population. In Salisbury, on the banks of the Little Para River, stormwater pooled on low-lying, clay-rich soils near the subdivision of Para Hills. With little support for developing this flood-prone area for housing, the local council reserved what became known as 'The Paddocks' as a wetland in 1975. Another followed as the council recognised that the reed beds slowed the flow of water and filtered out pollutants while attracting bird life and providing a space for recreation.[128]

Elsewhere in Adelaide, the pollution of the River Torrens and Patawalonga, as well as West Lakes and the Onkaparinga, continued to frustrate local residents.[129] As noted in Chapter 7, concerns about the pollution of urban waterways were shared among the five cities, with contamination increasingly recognised as a hydrological consequence of rapid urbanisation. 'Surface run-off', as researchers observed of Sydney's pollution in the mid-1970s,

> is a waste yield of city living that stems in large part from the comfort and convenience arising from the high quality, smooth, impermeable road surfaces on which we drive, the footpaths on which we walk, the gutters and drains that allow us all-weather access to our dwellings, places of work and recreation, and the parking lots where we park our motor vehicles.[130]

In the United States, with the *Clean Water Act* of 1972, the strategy of 'low impact development' was emerging to ameliorate the pollution of local waters and aquifers. Studies of US cities during the late 1970s and early 1980s confirmed that their impervious areas were changing the

[125] See for example Connell, *Water Politics*. [126] Kingsford et al., 'Engineering a crisis'.
[127] Miles, *Geology*.
[128] John Radcliffe et al., 'Fifty years of water sensitive urban design; Radbone and Hamnett, 'Making the most'.
[129] Pitrans, 'Stormwater management'. [130] Butlin, *Environmental Amenity*, 145.

hydrology and contamination of waterways, prompting the closer regulation of urban run-off.[131]

Since the sewage and stormwater systems of most Australian cities are not combined, as they are in the United States, the adoption of these North American approaches was neither straightforward nor desirable. An outbreak of algal blooms in Perth's Swan River in the early 1990s encouraged urban planners to formulate a strategy for 'water sensitive urban design' that might ameliorate the flows of phosphate and nitrate into the waterway. Focusing on the city's largest drainage catchment, the Bayswater Drain, planners understood their approach as part of a wider reappraisal of the city's relationship with the natural environment, as cities formed urban water catchments of their own. They sought a twenty-first century alternative to the nation's '19th century solutions to urban water management' that were characterised by a 'big pipes in – big pipes out' logic, which the Hawke government's Industry Commission had recently identified as economically restrictive.[132] An approach to urban design that would 'manage the water balance', 'maintain and where possible enhance water quality', 'encourage water conservation', and 'maintain water-related environmental and recreational opportunities' was advocated.[133]

Australian planners were also responding to growing interest in the recreational uses of the nation's urban waterways. Because most urban water catchments were closed to protect drinking water supplies, there was greater pressure on other 'water-based recreational resources'.[134] Adelaide's River Torrens Linear Park Scheme, begun in 1982, is a flood mitigation project that stretches nearly 50 km from the hills to the coast, serving as both a transport corridor for the O-Bahn busway and an urban park with a walking and cycling trail.[135] Deindustrialisation and a growing market for leisure activities also guided plans for the Penrith Lakes Scheme, the redevelopment of the open-cut sand and gravel quarries at Castlereagh, on Sydney's western outskirts, into a vast artificial waterscape of recreational and wetland lakes the size of Sydney Harbour.

131 Jones et al., 'Urban storm-water'; Radcliffe, 'History of water'.

132 Mike Mouritz, 'Towards sustainable solutions', *Urban Wastewater: A Lost Resource or an Opportunity, an AIUS, AWWA and WAWRC Seminar 17 August 1993, Perth* (East Perth: Australian Institute of Urban Studies, 1993), 165; Industry Commission, *Water Resources and Wastewater Disposal*, Inquiry Report No. 26 (Canberra: Australian Government Publishing Service, 1992).

133 Cited in Fletcher et al., 'SUDS, LIDS'.

134 Pigram, 'Outdoor recreation'.

135 Rolls, 'Stormwater management'.

Although quarrying in the area continued until 2020, a rowing lake was completed in time for the 2000 Olympics.[136]

Applications of water sensitive urban design principles were slow to emerge. For all Adelaide's efforts, the Patawalonga had in 1994 earned the unhappy distinction of being Australia's most polluted waterway – 'a cess pool', in the words of Prime Minister Keating.[137] Some critics attributed this sluggish progress to the corporatisation of the nation's water utilities. Hydrologist Peter Cullen speculated in the mid-1990s that these newly privatised bodies would rather focus on 'core business that generates a cash flow', echoing the cost-saving focus of Sydney's 1978 McKinsey report discussed in Chapter 7.[138] The fragmentation of urban water management was also an obstacle for systematic flood control and drainage: in Adelaide, thirty local councils were responsible for six or seven catchments in the mid-1990s, with almost none of their boundaries aligning with catchment areas.[139]

Although Adelaide was growing only slowly, a gradual increase in its population density reflected the more substantial and rapid urban consolidation underway in the other cities, especially Melbourne and Sydney.[140] Influenced by Commonwealth policy, this trend was shaping the hydrology of the nation's cities and their surrounding areas, as journalist Clem Lloyd observed in 1995: 'Urban tarmac ... is a superb catchment for capturing run-off and discharging it pell-mell to overloaded drainage and sewerage systems.'[141] Marine ecologists estimated that seagrass cover off the coast of Adelaide's Gulf St Vincent had declined from 80 per cent in 1949 to less than 30 per cent in 1993.[142] A four-year study had found that while Port Phillip Bay was in surprisingly good health, the flow of nitrogen from Melbourne's drains and waterways should be dramatically reduced.[143]

Subsequent regulatory frameworks and guidelines to attend to the engineering and environmental challenges of urban run-off dovetailed with Federal and State government approaches to the Millennium Drought.[144] Building on the inclusion of water-sensitive urban design

[136] Karskens, 'Water dreams'.

[137] Paul Keating, 'Statement by the Prime Minister: Patawalonga Clean-Up', *PM Transcripts*, 7 October 1994, https://pmtranscripts.pmc.gov.au/release/transcript-9371, accessed 7 April 2021.

[138] Cullen, 'The Cinderella', 15. [139] Rolls, 'Stormwater management'.

[140] Baker, Coffee, and Hugo, *Suburbanisation*. [141] Lloyd, 'Water, sewerage', 70.

[142] Edyvane, 'Issues', 65. [143] Melbourne Water, *Port Phillip*.

[144] See for example Lloyd, *Water Sensitive*.

principles in the 2004 intergovernmental National Water Initiative, a 2007 report to Prime Minister Howard argued:

> Water supplies to Australia's cities need to move from reliance on traditional sources to an efficient portfolio of water sources which can provide security through diversity. Like a share portfolio, flexible and cost-effective access will be underpinned by diversity, including centralised and decentralised water infrastructure.[145]

Prepared by a non-governmental advisory group, the report pointed to the possibility of using the city itself as a water catchment. Not simply a nuisance to be detained and treated for flood and pollution mitigation, stormwater could be captured, stored, and later harvested as a water resource.

At their meeting, members of this advisory group drank 'Recharge' bottled water, collected from an urban stormwater treatment scheme in Adelaide. The success of recent trials of extracting potable water from aquifers in which stormwater had been stored suggested that prospects of expanding this Parafield Gardens scheme looked bright.[146] Since the late 1980s, several South Australian municipalities had found that naturally brackish or saline aquifers could serve as underground 'tanks' in which stormwater could be treated and stored for later use. These ambitions soon fell afoul of politicking in the lead-up to the 2010 State election, after which the incumbent Labor government maintained its opposition to what it had mischievously described as a 'risky plan to inject road run-off straight into mains water system'.[147] More palatable has been the reuse of stormwater for irrigation and industrial uses, which is much cheaper than potable water supplies, as well as to service residential developments with dual water supply systems, with fifty aquifer recharge schemes in operation in 2017.[148]

Since the end of the Millennium Drought, Federal and State government funding has encouraged the further elaboration and application of the principles of water-sensitive urban design, with the aim of developing 'water-sensitive cities' in Australia and abroad. Given the path-dependent nature of urban water infrastructure, these design approaches have been

[145] Prime Minister's Science, Engineering, and Innovation Council, *Water for Our Cities: Building Resilience in a Climate of Uncertainty – a Report of the PMSEIC Working Group* (2007), 11.

[146] Dillon et al., 'A critical evaluation'.

[147] Michael Owen, 'CSIRO caught in storm over water', *Australian*, 11 February 2010.

[148] Krestschmer, *Managed Aquifer*.

FIGURE 8.3. Salisbury recycled water, Adelaide (photographer: Baden Myers).

more likely to be adopted in new developments, subdivisions, and housing estates, and in more piecemeal fashion in established suburbs.[149] Nearly two decades after Australia's first dual-reticulation urban housing project provided recycled water for gardens, toilet flushing, and car washing at Rouse Hill in north-west Sydney, the installation of separate pipes for potable and non-potable uses has since become mandatory in new housing subdivisions in most states, along with other water conservation infrastructure such as rainwater tanks.[150] Nevertheless, a 2015 Senate inquiry into the management of the nation's urban stormwater found that significant financial, research, and institutional barriers remain across the states and were hindering the wider use of stormwater in Australian cities.[151]

Over the past decade, researchers have also interpreted these principles as a means to cool Australia's warming cities. In addition to the likelihood of more intense rainfall events, Australian climate scientists predict

149 Barton and Argue, 'A review'. 150 Radcliffe et al., 'Water reuse'.

151 Senate Environment and Communications References Committee, *Stormwater Management*.

warmer temperatures and more frequent and hotter heatwaves for the nation's cities.[152] In Western Sydney, where the State government anticipates another million people will live by 2031, increasing summer heat and heat-absorbing surfaces could lead to temperatures as high as 50°C (122°F) by 2050.[153] As a local administrator warned in 2021, 'if the community doesn't have access to things to make themselves cool we effectively have a very large oven occurring across the Western Sydney Plains'.[154] During the Millennium Drought, local councils and schools had made the most of whatever water they could to irrigate parks and sports ovals. In Melbourne, ageing and drought-affected tree species such as elms and poplars are gradually being replaced with drought-tolerant alternatives.[155] Aside from avoiding the unsightliness of the deterioration of local vegetation, climate researchers suggest that harvesting stormwater could ensure the longevity of the grassed areas and tree canopies that could help to keep Australia's growing cities cool.[156]

WEALTH FROM WASTE

Drier futures for the growing southern Australian capital cities suggest that the economic and environmental value of wastewater will also increase as alternatives become relatively more expensive and inaccessible. Western cultural mores and Chadwickian sanitary reform have long combined to foster systems for the rapid, permanent, and sterile expulsion of human effluent from human life.[157] By physically displacing this waste, these processes historically detached human consumption from the visceral materiality of its resulting excrement.[158] The infrastructures of wastewater reuse facilitate the physical reconnection of these components of the urban metabolism in productive and even palatable ways. Waste

[152] Lewis, King, and Mitchell, 'Australia's unprecedented'.

[153] Yasmin Parry, 'How Western Sydney is tackling the mysterious "heat island" effect behind rising temperatures', *ABC News*, 1 March 2018, www.abc.net.au/news/2018-03-01/how-western-sydney-is-tackling-the-heat-island-effect/9361156, accessed 7 April 2021.

[154] James Purtill, 'Heatwaves may mean Sydney is too hot for people to live in "within decades"', *ABC Science*, 24 January 2021, www.abc.net.au/news/science/2021-01-24/heatwaves-sydney-uninhabitable-climate-change-urban-planning/12993580, accessed 7 April 2021.

[155] City of Melbourne, *Urban Forest Strategy: Making a Great City Greener, 2012–2032* (Melbourne: City of Melbourne, 2011).

[156] Coutts et al., 'Watering our cities'. [157] Jewitt, 'Geographies of shit'.

[158] See for example Hawkins, 'Waste in Sydney'; Moore, 'Garbage matters'.

may not be expelled, but its presence in treated form retains the 'literal and moral distance from bodily waste' that city dwellers prefer.[159] The waste component of the city's metabolism remains invisible, while potable water flows unabated.

In response to the deteriorating condition of Perth's groundwater reserves and ongoing decline in rainfall, the Western Australian government began a trial in 2011 to replenish the Gnangara Mound with treated wastewater.[160] Satisfied with the results, the Water Corporation announced in early 2013 that it would add this recycled water to the city's drinking water supplies.[161] Mirroring similar schemes in Monterey, California, and Changwon, South Korea, municipal wastewater undergoes several stages of treatment before its injection into aquifers, where it mixes with the existing groundwater before being extracted for conventional treatment and distribution.[162] Perth is the only city of the five to pursue this approach to supplement drinking water supplies; in Adelaide and Brisbane, wastewater is treated and reused only for industry, irrigation, and energy production. The Perth groundwater 'phase' may have aided the popular acceptance of the program, as groundwater has for decades been a significant source of water for the suburbs, albeit outdoors. Perth's approach has so far avoided the 'yuck factor' associated with other efforts to establish water recycling schemes in Australia and elsewhere. Residents of drought-stricken Toowoomba in regional Queensland, for instance, rejected a 2006 proposal to add recycled water to their city's water supply, which they claimed would turn their town into 'Poowoomba'. Consequently, Toowoomba residents subsidise the cost of pumping water up the Great Dividing Range from Wivenhoe to supplement their local supplies.[163]

Perth's groundwater buffer may prove beneficial for not only the city's sanitary needs but also the wider environment and its cultural significance. Pumping the treated wastewater into the aquifer could improve the condition of urban wetlands, while ameliorating salt water intrusion from both estuaries and the ocean.[164] During the drought of the early 2000s, the State's Environmental Protection Authority had noted that observed

[159] Hawkins, 'Waste in Sydney', 350.
[160] Daniel Mercer, 'Sharp fall in Gnangara water levels', *West Australian*, 10 June 2011.
[161] M. Hopkin, 'Recycled waste water gets go-ahead for Perth', *WA Today*, 1 August 2013.
[162] Khan, *Drinking Water*.
[163] Hurlimann and Dolnicar, 'When public opposition'. See also Spearritt, 'The water crisis'.
[164] McFarlane, 'Will Perth', 209–38.

water levels in local caves had declined to such an extent that 'some loss of species is likely to have occurred'.[165] Some wetlands had also experienced fires and greater acidity due to the dry conditions.[166] The deterioration of Perth's groundwater reserves and wetlands also had cultural implications for many local Indigenous people. Where Noongar had once been able to rely on soaks at swamps and other sources for fresh water, these were now dry.[167] Aboriginal people told anthropologists that the declining water table was affecting the growth of medicinal plants and that much of the local fauna was no longer present. The ailing health of these wetlands had serious consequences for Aborigines: as one Noongar man pointed out, 'without water there wouldn't be any Aboriginal spiritual life'.[168]

Although the Water Corporation was focused on the sustainability of the city's water supplies, elsewhere Perth's wetlands have undergone a significant ecological and cultural reappraisal. In the 1950s, Gordon Stephenson and Alistair Hepburn, the planners charged with developing a metropolitan plan for Perth and Fremantle, identified North Lake and Bibra Lake as important regional public open space. They also planned for a highway to run between the lakes, taking freight through the southern suburbs of Perth to the port at Fremantle. At the time, this proposal was not controversial – there was no public consultation on the plan nor any environmental review process. Highways were also understood to be an essential part of the efficient, modern city of the future and not necessarily incompatible with public enjoyment of the area.[169] The rise of environmentalism in the 1970s fostered public appreciation of the aesthetic and ecological values of wetlands, as well as problems associated with major roads. A proposal in the mid-1980s to extend Farrington Road across the northern end of North Lake became the site of a heated environmental conflict. The Main Roads Department pressured the Environmental Protection Authority (EPA) to approve its plans, and weeks later the bulldozer moved in. The local community and students of nearby Murdoch University responded by standing knee-deep in the wetland in front of the bulldozer during the day and removing surveyors' star pickets at night.[170]

165 Water and Rivers Commission, *Environmental Management*, 6.
166 Water and Rivers Commission, *Environmental Management*, 6.
167 McDonald, Coldrick, and Christensen, 'The green frog, 69.
168 Cited in McDonald, Coldrick, and Christensen, 'The green frog, 66.
169 Gaynor, Jennings, and Newman, 'A history of Roe 8', 16.
170 Gaynor, Jennings, and Newman, 'A history of Roe 8', 18–19.

This was not the first time a Perth wetland had been the subject of environmental conflict: an ultimately successful campaign to save Star Swamp in Perth's northern suburbs from obliteration by residential development ran from 1976 to 1985. However, the intensity of the Farrington Road blockade took the government by surprise, crystallising the depth of public feeling about wetlands and opposition to the 'develop at all costs' paradigm of preceding decades. The road was built, though the protesters gained some concessions resulting in a degree of preservation, subsequently forming the Wetlands Conservation Society to care and campaign for Perth wetlands. In the wake of the dispute, the Society extracted a government commitment to establish the Beeliar Regional Park encompassing the major wetland chains in Perth's southern suburbs, including North Lake and Bibra Lake. Premier Brian Burke, hoping to avoid a sequel to the Farrington Road conflict, also directed the State Planning Commission and Main Roads to explore alternative routes for the proposed Roe Highway extension – Roe 8 – that would pass through the wetlands. While the working group and EPA agreed that the highway should not be built through the wetlands, the road reserve remained in place.

When the Beeliar Regional Park was established in 1997, the Roe 8 reserve was excluded from it. It remained a live issue, with 1500 people attending a rally against Roe 8 in the lead-up to the 2001 election. Colin Barnett's Liberal–National Coalition government, elected seven years later, was committed to reviving Roe 8, and a proposal was developed and submitted to the EPA for approval. During the Public Environmental Review, over 3000 submissions were received, the vast majority of which opposed the proposed route through the wetlands. In spite of its policy that offsets should not be used in the case of critical environmental assets, the EPA approved the proposal with conditions and offsets, thereby reversing its long-established position that Roe 8 was environmentally unacceptable.[171]

The proposal also raised significant Aboriginal heritage issues. The Beeliar wetlands were important Whadjuk Noongar places: as major camping places, they are sites of archaeological significance, and as significant wetlands with deep water, they are mythological sites associated with the Waugal. This is also the case at sites such as Lake Claremont or 'Butler's Swamp', Perry Lakes in Floreat, Jolimont Swamp, Lake Monger,

[171] Gaynor, Jennings, and Newman, 'A history of Roe 8', 25–26.

Hyde Park, Dogswamp, and Lake Gnangara.[172] While the committee that assesses applications to damage or destroy Aboriginal heritage sites recommended against consent to proceed with Roe 8 in 2013, a cursory survey by Department of Aboriginal Affairs staff in 2014 saw the site deregistered in 2015, the same year in which the project also received Commonwealth funding and environmental approval. A legal battle in the state courts ensued, with the Save Beeliar Wetlands group seeking judicial review of the EPA's decision; the state's Supreme Court found that the EPA assessment was invalid and the environmental approval was void. This ruling was then overturned on appeal, and work on the highway commenced in late 2016.[173]

What followed was one of the largest and most high-profile community protest actions around urban wetlands in Australia to date. Fewer than three months from an election in which the ALP committed to halting the project, contractors began fencing and clearing the reserve. In response, a diverse, lively, and committed campaign mobilised thousands of people.[174] Wetland protectors – many of them Aboriginal – camped near the site, monitoring works for breaches of approval conditions. Supporters from across the social spectrum were prepared to risk time, money, and reputation to protect the wetlands, together demonstrating the forms of group organisation that Elinor Ostrom observed in resistance to top-down, politically imposed directives. This outpouring of love for the wetlands from both Indigenous and non-Indigenous peoples, and anger at the government's dogged pursuit of a flawed policy, was vindicated when the ALP, led by Mark McGowan, won the March 2017 election and stopped work on the project.

The election of the McGowan government also brought a brief reprieve to the pine plantations of the Gnangara Mound and their animal residents. Since the 1940s, large flocks of Carnaby's black cockatoo have been attracted to the pine plantations established above the Mound two decades earlier. Once seasonal visitors to the area, the accelerated clearing of their historic habitat in what is now the state's wheatbelt led some birds to take up residence in the plantations. As Perth and its dependence on groundwater grew during the 1970s, these pines would soon be regarded as competition for human water supplies. Despite the declaration of the species as endangered in the mid-1980s, the cockatoos factored little in

172 O'Connor, Bodney, and Little, *Preliminary Report*, 13–20, 25–35, 38–39.
173 Gaynor, Jennings, and Newman, 'A history of Roe 8', 8–9.
174 Whish-Wilson et al., 'A twenty-first'.

the decision of Charles Court's Liberal government to make an agreement with a company to harvest pines for the production of laminated veneer lumber.[175] After the pines were cut back to just over 20 per cent of their former area, from 23,000 hectares to 5000 hectares, the McGowan government paused the area's harvest. The harvest resumed in mid-2019. Birders estimate that the cockatoo's Perth population has declined by 35 per cent since 2010, while local volunteers have begun to plant native trees to supplement the growing shortfall of food.[176]

For all the potential benefits of wastewater recycling, from enhancing water supplies to ecological restoration, the strategy is not without its problems. Chief among them is the perpetuation of existing large-scale, centralised approaches to water management, which help to facilitate inconspicuous consumption as a matter of convenience.[177] The hidden nature of the wastewater recycling process might continue to insulate water users from the environmental toll of their behaviours. There are also implications for the climate. Although this form of water resource development might ameliorate some impacts of urban development and water use on local aquifers, groundwater replenishment has a sizeable carbon footprint.[178] At present, the Perth scheme exceeds the carbon footprint of the city's two desalination plants, the emissions of which are offset by nearby wind and solar energy plants. At time of writing, the Western Australian government is building a renewable energy generator (which will use biogas, a by-product of the wastewater treatment process) to supply the needs of an expanded wastewater recycling scheme.[179]

[175] O'Gorman and Gaynor, 'More-than-human'.
[176] Peck, Barrett, and Williams, *The 2019 Great Cocky Count*; Tyler Brown and Lucy Jarvis, 'Crowdfunded trees planted for cockatoo food', *Wanneroo Times*, 20 July 2020.
[177] Sofoulis, 'Big water'; Shove, *Comfort*.
[178] Simms, Hamilton, and Biswas, 'Carbon footprint'.
[179] Water Corporation, 'Renewable energy'.

9

Epilogue

In June 2019, bushfires began in Queensland. At the end of that 'Black Summer', across Australia fires had killed thirty-three people, destroyed 3094 houses, burnt 17 million hectares of land, and killed more than an estimated one billion animals.[1] On 18 December, the country experienced its hottest day on record, with the national average temperature reaching 41.9° C.[2] In the same month, Sydney had level 2 water restrictions that were lifted when the rains that helped to extinguish the bushfires arrived in March.[3] In January 2020, Australia experienced its first COVID-19 case.

In October 2020, hailstones caused over a billion dollars damage to homes in Brisbane, only to be followed in March 2021 by another hailstorm devastating 130 houses.[4] Two days later, people living in

[1] Parliament of Australia, *2019–2020 Australian Bushfires – Frequently Asked Questions: A Quick Guide*, Research Paper Series 2019–2020, 20, https://parlinfo.aph.gov.au/parlInfo/download/library/prspub/7234762/upload_binary/7234762.pdf, accessed 19 April 2021.

[2] Graeme Readfearn, 'Australia records its hottest day ever – one day after previous record', *Guardian*, 19 December 2019, www.theguardian.com/australia-news/2019/dec/19/419c-australia-records-hottest-ever-day-one-day-after-previous-record, accessed 19 April 2021.

[3] Level 2 water restrictions restrict people from watering their gardens to the hours before 10 am or after 4 pm with a watering can or bucket. Even smart and drip irrigation systems may only be used for 15 minutes in the same period. Cars may only be washed by bucket or in a commercial car wash. Hosing of hard surfaces is not permitted, except in an emergency. Swimming pools may only be filled if a permit is issued, while topping up pools and spas is limited to 15 minutes a day. Households may be fined $220 for breaching the restrictions, while businesses face fines of $550. *Sydney Morning Herald*, 21 November 2019.

[4] 'South East Queensland storms wreak havoc on areas still recovering from previous storm damage', www.abc.net.au/news/2021-03-03/bom-storms-brisbane-ipswich-communities-recovering-springfield/13210128, accessed 19 April 2021.

Queensland's south-east were warned of impending water restrictions, as the region's dams were at their lowest levels since the Millennium Drought ended in 2009. Authorities were initiating the start-up of a controversial scheme to pump treated wastewater into the water supply.[5] Less than three weeks later, severe thunderstorms lashed most of the east coast of Australia. Intense rainfall, in some areas over 100 mm falling in an hour, caused flash flooding throughout the Gold Coast and hinterland. Further south, the New South Wales Bureau of Meteorology issued flood warnings for thirty-three of the forty river catchments that flowed to the east coast. The three largest, the Hawkesbury-Nepean, Clarence, and Hunter rivers, were all flooded. Tens of thousands of people were ordered to evacuate from the Hawkesbury-Nepean River area in Sydney.[6]

At the time of writing, the spread of COVID-19 continues to test the capacity of public health systems around the world. Although the pandemic's toll on Australians has been comparably mild, travel restrictions across suburban, state, and international borders, as well as lockdowns of varying duration in all five capital cities have raised questions as to the form, density, and desirability of Australia's capital cities. Near universal access to clean water ensured Australians could follow World Health Organization guidelines to wash their hands often with soap and water to help prevent the spread of the disease. As part of an ongoing public health program of wastewater monitoring, health authorities test sewage for traces of the virus, which can help to detect the spread of the disease and inform public health responses.[7]

Australia's capital cities are all located on freshwater rivers. Aboriginal peoples well understood the importance of access to adequate water and had inhabited these spaces for millennia, taking advantage of the fresh water, plentiful plant life, native animals, and the comforts of fertile locations. They had built a sustainable culture around a deep understanding of, and integration with, the non-human world that enabled them to adapt to long-term environmental changes and preserve the environmental knowledge acquired over many generations. When the European

[5] Felicity Caldwell, 'Water grid falls to lowest level after summer since Millennium Drought', *Brisbane Times*, 4 March 2021, www.brisbanetimes.com.au/politics/queensland/water-grid-falls-to-lowest-level-after-summer-since-millennium-drought-20210304-p577pw.html, accessed 19 April 2021.

[6] 'NSW floods unmatched in scale and rainfall, but history shows there have been worse', www.abc.net.au/news/2021-03-26/nsw-floods-unmatched-in-scale-and-rain-but-there-have-been-worse/100026690, accessed 19 April 2021.

[7] Donner et al., 'Wastewater monitoring for SARS-CoV-2'.

colonisers arrived, they were attracted to the same sites that the Aboriginal people had long inhabited. As the invaders stayed and claimed the lands and waters, conflict with the resident Aboriginal population almost invariably ensued.

The newcomers' approach was based on agricultural farming: a static process of occupation advanced through land ownership and control. Unlike the symbiotic and communal relationship between Aboriginal peoples and Country, whites pursued a faith in settler capitalism with its accompanying racial logic of land-ownership that justified Indigenous dispossession. In some cases, the settlers' reservation of riverbanks and wetlands for access to water enabled Aboriginal people to live near and maintain their relationship with important sites. Occasionally this occupation persisted well into the twentieth century, even as settlers degraded rivers through pollution and river-bank clearing and contained or drained wetlands. In many cases, however, Aboriginal people's links to the most fertile areas were irretrievably destroyed.

The colonists also brought with them an aggressive determination to found and build societies that would be the envy of the old world. On seemingly 'untouched' frontiers, they were resolute in their efforts to dominate the original inhabitants and the 'natural' environment and so create prosperous societies for themselves and their children. To achieve this, the colonists set about planning, building, and controlling their townships and exploring the hinterland to begin the transition from what they perceived as untamed wilderness to land that was familiar and controlled. Critical to this process was the identification and capture of freshwater. Through damming sources and constructing networks of pipes from the mid-nineteenth century, each town gradually built systems to ensure its reliable delivery and distribution.

The process of construction, expansion, and delivery of water services in urban settings was not straightforward. Not only did physical and geographical barriers need to be overcome, but in several cases, social, institutional, and financial constraints so delayed progress that crises emerged. As populations grew and demand outstripped supply, it often took several years, and not infrequently the emergence of a health crisis, to spur action. In the nineteenth century, as settler understanding of diseases shifted from theories blaming poverty and miasma to more bacteriological approaches, sanitary techniques such as cesspits came to be seen as part of the problem. Overcoming the legacies of previous decisions while also keeping pace with expanding urban populations often exceeded the capacity of governments and their departments.

Constructing infrastructure to deliver water was usually given a higher priority than building systems to remove sewage. Early solutions, such as sewage farms, often proved inadequate, more for reasons of aesthetics and lack of planning or political will to upgrade them than because they were necessarily a health hazard. Of the five cities, only Melbourne continued with sewage farming into the twenty-first century. The replacement technology in other cities, sewerage systems that processed and aerated sewage before 'disposal', took over a century to develop beyond practices that did more than dump minimally 'processed' waste into the ocean. Even today, most stormwater still goes into waterways without any treatment at all.

At Federation in 1901, Australia was one of the most urbanised nations in the world. While the cities of Sydney and Melbourne were comparable in size to several other large Western cities and their water networks in advance of many, they had still only recently achieved success in supplying the larger part of their populations with access to at least an outside tap. While the dangers of poor sewerage systems and inadequate drainage were by now well known, most suburban households still only had access to outside toilets serviced by a pit, pan, or septic tank. The sewerage networks were concentrated in suburbs with easier topography and households of middle- and upper-class income. Adelaide, Brisbane, and Perth, all smaller cities, lagged behind, as the constraints of slower financial growth and later development held them back. Nonetheless, for most Australians, access to fresh water was much less of a problem than it had been a generation earlier.

While capital works were put on hold during both world wars, State governments and their water departments continued to expand the water storage systems available to each city, building new dams and constructing thousands of kilometres of pipes. With each major drought, water rationing spurred calls from citizens in Adelaide, Perth, Melbourne, and Sydney for increased storages and more reliable systems. In Brisbane, flooding triggered calls for dams to serve as protective infrastructure. The technical and management skills needed to solve these challenges should not be underestimated. On the driest inhabited continent on the planet, Australians built huge urban networks that, for the most part, delivered safe, reliable water and effectively removed human waste. This process of building the nation and creating liveable cities was underpinned by the activities of state governments. Urban planning, however, only occasionally included designs that accepted environmental conditions; long-term plans were those that looked twenty or thirty years into the future. Rarely, if ever, did planners see insurmountable constraints to continued development.

As suburbs expanded, networked electricity and transport systems were accompanied by rising demand for consumer goods, increased convenience, expectations of reliable water supply, and safe waste disposal. Household gardens became greener and more elaborate as water availability increased, even if a key objective in many poorer suburbs was to grow vines over the ageing outside toilet and infrequently used rainwater tank. Inside, bathrooms and kitchens with integrated hot and cold running water were still a measure of social status and prestige, rarely found in often dilapidated rental dwellings. While some enjoyed the opportunity to soften harsh summers with plentiful water, and newer housing designs progressively integrated toilets and bathrooms under the main roof, many others endured limited or ad-hoc facilities supplying water to a 'lean-to' built on the back of old housing stock.

The rapid expansion in immigration and industry after World War II saw most cities' water networks struggling to keep up with the increasing demands of the growing suburbs. Public water authorities and government departments responded by constructing public works, but financial and political constraints meant they were always 'behind'. Householders in some newer post-war suburbs met their own needs for wastewater disposal – and in some cases built their own street drainage and walkways. For the next quarter of a century, however, access to unlimited quantities of cheap, reliable, clean drinking water and efficient, safe waste disposal was a basic assumption of most urban dwellers, secure in the knowledge that government engineers would overcome any short-term problems. Inside and outside the home, in private and public gardens, in parks and at ovals and in public spaces, an abundance of water was regarded as both fundamental and a right. Urban residents mostly assumed their city was 'drought proof' and supply assured. Only occasionally, when nature sent extreme and 'unusual' droughts, were these expectations challenged by water restrictions: restrictions that could be overcome through more and larger dams. Consequently, urban demand for water steadily increased.

This pattern of growth under-estimated the environmental limits of continued expansion. By the second half of the twentieth century, the strain was showing. In Adelaide, there were no more sites for major dam storages. In Perth, the last of its large dams was built, and the desalination of seawater was under consideration. In Sydney and Melbourne, existing waste disposal systems were creating unacceptable levels of sea pollution.

In Brisbane, the significant expansion of suburbs across flood zones, and the increased density of buildings along the Brisbane River, awaited the next 'big one'. As populations and cities increased, so did the demand for water and sewerage facilities. While the capacity of the environment to absorb such demands decreased, the financial, social, and energy costs of maintaining supply continued to increase.

To lower aggregate demand for services while still increasing the cities' populations required individuals, households, and businesses to use less water and create less waste. The solution that was chosen to change behaviour was to shift citizens away from expecting a right to cheap water and sewerage towards expecting to pay for their use. From the 1970s, and over the next quarter of a century, this process entailed changes to government institutions (privatising or at least corporatising water and waste services); service charges (paying for the amount of water used and, to a lesser extent, the amount of waste removed); changes to infrastructure (from improved networks to subsiding rainwater tanks and reduced-flow shower heads), and educating citizens to become more frugal with their water use (from introducing permanent water restrictions to promoting native gardens and water tanks).

The costs and benefits of these changes are still emerging. While the amount of water consumed per person has decreased in the mainland cities, aggregate demand has grown with increased populations. Socio-economic inequalities manifest when the gardens of blue-collar households suffer when water restrictions apply, while 'private bore' (or tank water) signs proliferate in elite, greener, and geographically well-situated suburbs. For the most part, sewage, although more safely treated than half a century ago, still involves pumping millions of litres of effluent into the sea near each mainland city – though Melbourne's sewage farm at Werribee also serves as a major wetland and bird sanctuary. While several plants treat waste to a 'tertiary' level, some in Sydney still pump untreated sewage out to sea. Responsibility for systematic health standards now include a commercial consideration rather than an absolute standard. 'Big Water' (corporations) rather than 'big government' (state departments) now control urban water networks. Reputation, as much as reliability, determines how well providers can operate to supply the market. While the costs of supply and removing water are rising, the limits of individuals to pay is likely to eventually erode commercial motivations. Environmental costs, particularly if outside market mechanisms, remain problems rarely addressed by private providers.

Nor have the challenges become easier. Where once supplying water to a city was an engineering and planning challenge because of lack of finance and capital infrastructure, the task is now to overcome path-dependent barriers caused by existing population centres, installed structures, and prevailing physical and social networks and cultures of water use. Costly projects, such as 'retro-fitting' networks of storm water collectors with connected wetlands or building distribution networks for recycled water fall increasingly on State and local government, while income-generating activities remain with the private sector. Consumers' willingness to pay for change, especially where the improvement is not easily seen (as in the environment) or appears to accrue to others, also remains a significant political and social constraint. Changing behaviour, such as encouraging the consumption of treated effluent, remains a political and social challenge.

While the challenges of the past were substantial, the challenges of the future are no less daunting. The impact of climate change not only threatens the amount of rainfall for the cities of Perth and Adelaide but also increasing volatility in weather events threatens capital infrastructure and water networks in Brisbane and Sydney. The threat (and in some cases reality) of less rain has seen desalination plants built in Perth, Adelaide, Melbourne, Sydney, and the Gold Coast, south of Brisbane. Frequently built in times of high anxiety, no desalination plant was subject to a rigorous cost–benefit analysis, about either energy costs or environmental impacts. The threat of increased extreme rainfall events has seen calls for higher (and lower) dam walls in Sydney and Brisbane, while such extreme events also damage cities' treatment plants and waste disposal systems. In the long term, rising sea levels may not only threaten underground aquifers, as in Perth but also, given their locations, the capacity of most wastewater treatment plants in all the mainland cities. Networks that have been constructed as permanent solutions to water distribution will need to be modified and, in some cases, moved, as climate-related changes shift population centres. New networks that are sensitive to local conditions and better integrate management of stormwater and treated wastewater with fresh and desalinated water remain under-developed.

For 230-odd years, Australians have demonstrated their ability to adapt to change. Urban Australians have built, and rebuilt, water and waste infrastructure as settler understandings of local climates, sanitation, engineering, and science changed – and as urban populations grew. In the nineteenth century, this meant replacing cesspits with sewerage networks or, later, small earth dams with much larger reservoirs based on significant engineering. From the late nineteenth century, every metropolitan capital chose to invest in centralised water-borne sewage systems, networks that often failed to keep pace with urban growth. Water closets were favoured over earth closet technology – a preference that is today being questioned in the face of rising environmental awareness. In the early twenty-first century, an Australian government website providing information on sustainable housing recommends waterless toilets as 'a genuine, minimum energy, on-site alternative to centralised reticulated systems that transport the problem downstream' that could be 'an elegant addition to a modern bathroom'.[8] We have begun increasing our capacity to harvest water that was previously wasted or lost and to create fresh water from the sea. Individual products, buildings, new streets, and whole suburbs are now being designed to reduce water loss and maximise water collection and reuse. Wetlands, highly valued by Aboriginal peoples, that were once drained for reasons of sanitation or urban development are now in some places being partially restored as a result of their value for water conservation and stormwater storage being recognised.

Our understanding of the value of water in cities has also changed – and continues to evolve. The 'First Fleeters' were, in historian and broadcaster Michael Cathcart's evocative phrase, 'wet-country' people.[9] They understood the importance of water but initially assumed they could build their settlement on British patterns of its supply and demand, presuming access to water was cheap. Over subsequent generations, the scientific knowledge, engineering techniques, governance institutions, and social values that were either received from abroad or were modelled on them had to be absorbed and adapted to Australian conditions. Water was valued as precious, but settler Australians expected it to be supplied cheaply and reliably, unhindered by the effects of the continent's extreme climate variability. Writing during the Millennium Drought, Jay Arthur wondered:

> What if we dropped the word 'drought' from Australian English and used 'El Nino season' to refer to the phase of normal climatic conditions we presently call 'drought'? What if we abandoned the terms 'summer' and 'winter' and developed new terms which speak for southern Australia as 'The Dry' and 'The Wet' speak

[8] 'Waterless toilets', YourHome: Australia's guide to environmentally sustainable homes, www.yourhome.gov.au/water/waterless-toilets, accessed 21 April 2021.

[9] Cathcart, *The Water Dreamers*, 8.

of the seasons of tropical Australia? What if we had words, as yet unformed in English, which could talk about the 'unreliable' climate with terms of knowing?[10]

To meet the expectations of settler Australians in the growing coastal capitals, massive reservoirs and treatment plants and thousands of kilometres of pipes were planned and installed. For many years, the success of this approach was evident in city fountains, swimming pools, parks, and gardens. Socially, a lush green garden and, more recently, a personal swimming pool were relatively affordable for many. This is also beginning to change as people learn the true cost of such display and the risk of dependence on single sources. Economic incentives have helped curb demand, while household and product design reduce water use. Although some still view a lush European-style garden as a sign of social standing, to others a truly 'green' garden is one that is sustainable and conserves water. While 'drought-proofing' a city has been the stated aim of much recent research, many Australians have begun to realise that such a goal cannot be achieved only through new technology and infrastructure; how we use, save, and recycle what nature provides is also critical. Diversifying water sources, reducing the risks of over-reliance on single sources, and accommodating fundamental consumption needs while also modifying water use are among a range of future strategies to ensure ongoing and sustainable water supply and waste disposal in Australian cities.[11]

Urban Australians have been slow to change their use of water and the infrastructure built to move it. Growing city populations, higher-density urban settings, and, now, greater variability in weather events are increasing the pressure to adapt to change. The capacity of political and governance systems to solve wicked problems associated with water use through top-down directives is limited by path-dependent effects, from past infrastructure choices to widespread beliefs about water use and individual rights. Local environmental action, based on the empowerment of informed citizens in a liberal democracy, and respecting and learning from Indigenous perspectives on water management may be one approach to dealing with current and future problems facing urban Australians.

Over 65,000 years, generations of Indigenous Australians proved their resilience to the continent's extremes. They successfully responded to climatic and environmental changes and solved ever-pressing problems

[10] Arthur, *The Default Country*, 181.

[11] Leroux, Martin, and Zheng, 'Addressing water shortages by force of habit'.

of accessing water, disposing of waste, and living sustainably on Country. A little over two centuries ago, Britons with a completely different culture, values, technology, and perspective, and with a limited knowledge of its environment and constraints, claimed possession of Australia. These same two centuries have been the period of fastest global growth in all human history – as well as the period of most rapid environmental destruction.

More than 15 million people now live in Australia's five largest cities, 60 per cent of the national population. The problem of suppling clean safe water and removing waste efficiently, safely, equitably, and sustainably is an ongoing challenge for both current and future generations. If there is a lesson to be learnt from our past, it is that all choices have consequences. To ensure that the most precious of resources – clean, adequate water – can be accessed by all in the future, and waste safely removed, will require, regardless of the level of our sophisticated systems, living within nature's constraints. This is something that the original inhabitants demonstrated for millennia and that the newcomers are only belatedly learning.

References

Abbott, Carl, *How Cities Won the West: Four Centuries of Urban Change in Western North America* (Albuquerque: University of New Mexico Press, 2008).

Abelson, Peter W., *Privatisation: An Australian Perspective* (Sydney: Australian Professional Publications, 1987).

Acemoglu, Daron, and Robinson, James A., *Why Nations Fail: The Origins of Power, Prosperity, and Poverty* (London: Profile, 2012).

Acemoglu, Daron, Johnson, Simon, and Robinson, James A., 'The colonial origins of comparative development: An empirical investigation', *American Economic Review* 91 (2001): 1369–401.

Aird, W. V., *The Water Supply, Sewerage and Drainage of Sydney* (Sydney: Metropolitan Water Sewerage Drainage Board, 1961).

Allen, A. D., 'The Swan coastal plain: Hydrogeology of superficial formations', in B. A. Carbon (ed.), *Groundwater Resources of the Swan Coastal Plain* (Perth: Commonwealth Scientific and Industrial Research Organisation, Division of Land Resource Management, 1976), 12–31.

Allen, Michelle, *Cleansing the City: Sanitary Geographies in Victorian London* (Athens: Ohio University Press, 2007).

Allon, Fiona, *Renovation Nation: Our Obsession with Home* (Sydney: New South, 2008).

Alonso, A. D., 'How Australian hospitality operations view water consumption and water conservation: An exploratory study', *Journal of Hospitality and Leisure Marketing* 17 (2008): 354–72.

Anderson, Deb, *Endurance: Australian Stories of Drought* (Melbourne: CSIRO Publishing, 2014).

Anderson, Kay J., 'Place narratives and the origins of inner Sydney's Aboriginal settlement, 1972–73', *Journal of Historical Geography* 19 (1993): 314–35.

Anderson, Terry L., 'If Hayek and Coase were environmentalists: Linking economics and ecology', *Supreme Court Economic Review* 23 (2015): 121–40.

Anon., 'Obituary. William Clark, 1821–1880', *Minutes of the Proceedings of the Institution of Civil Engineers* 63 (1881): 301–8.

Ansara, Martha (ed.), *Always Was, Always Will Be: The Sacred Grounds of the Waugal, Kings Park, Perth, WA – the Swan Brewery Dispute* (Sydney: M. Ansara, 1989).

Armanasco, Peter, 'State position paper: Western Australia', in *Proceedings of the National Workshop on Urban Water Demand Management* (Perth: Western Australian Water Resources Council; Australian Water Resources Council; CSIRO (Australia) Division of Groundwater Research, 1987).

Arthur, Jay, 'The eighth day of creation', *Journal of Australian Studies* 23 (1999): 65–74.

The Default Country: A Lexical Cartography of Twentieth-Century Australia (Sydney: University of New South Wales Press, 2003).

Ashton, Paul, Cornwall, Jennifer, and Salt, Annette, *Sutherland Shire: A History* (Sydney: University of New South Wales Press, 2006).

Attenbrow, V., *Sydney's Aboriginal Past: Investigating the Archaeological and Historical Records*, 2nd ed. (Sydney: University of New South Wales Press, 2010).

Auditor-General of Victoria, *Building Better Cities: A Joint Government Approach to Urban Development* (Melbourne: Victorian Government Printer, 1996).

Australia, *Report from the Senate Select Committee on Water Pollution* (Canberra: Commonwealth Government Printing Office, 1970).

Australia Industry Commission, *Water Resources and Waste Water Disposal Report*, no. 26 (Canberra: Industry Commission, 1992).

Australia Productivity Commission, *Towards Urban Water Reform: A Discussion Paper* (Melbourne: Productivity Commission, 2008).

Australian Academy of Technological Sciences and Engineering (ATSE), *Perth's Water Balance: The Way Forward* (Melbourne: ATSE, 2002).

Australian Bureau of Meteorology, *Record-Breaking La Niña Events* (Canberra: Bureau of Meteorology, 2012).

Australian Bureau of Statistics, *Apartment Living* (Canberra: Australian Government Printing Service, 2016).

Australia's Environment: Issues and Trends (Canberra: Australian Government Printing Service, 2006).

Drought Drives Down Water Consumption (Canberra: Australian Government Printing Service, 2006).

Environmental Issues: People's Views and Practices (Canberra: Australian Government Printing Service, 1994).

Population Change: States and Territories (Canberra: Australian Government Printing Service, 2007).

Water Choices of Perth Households, Western Australian Statistical Indicators, 2010, no. 1367.5 (Canberra: Australian Government Printing Service, 2010).

Water Conservation: Water Conservation Devices Used (Canberra: Australian Government Printing Service, 2006).

Water Conservation Inside the Dwelling (Canberra: Australian Government Printing Service, 2007).

Water Consumption Down, Water Prices Up (Canberra: Australian Government Printing Service, 2010).

Water Supply in South Australia (Canberra: Australian Government Printing Service, 2008).

Australian Water Association, *50 years of Water in Australia 1962–2012*. https://f.hubspotusercontent30.net/hubfs/14568786/General%20File%20Uploads/PDFs/Others/AWA50thBooklet.pdf?__hstc=61638387.89b5ec8be5417b86979f6b36e3af92a5.1640930530133.1640930530133.1640930530133.1&__hssc=61638387.1.1640930530133&__hsfp=3234687218&hsCtaTracking=593d06d2-cfc4-48ae-9098-25b3f0593c65%7C181b9d83-0353-474a-b68a-bfe7363b3023.

Baker, Emma, Coffee, Neil, and Hugo, Graeme, *Suburbanisation vs Reurbanisation: Population Distribution Changes in Australian Cities, Australian State of the Environment Second Technical Paper Series (Human Settlements)*, Series 2 (Canberra: Department of the Environment and Heritage, 2000).

Bakker, Karen, 'Paying for water: Water pricing and equity in England and Wales', *Transactions of the Institute of British Geographers* 26 (2001): 143–63.

Baldwin, Peter, *Contagion and the State in Europe, 1830–1930* (Cambridge: Cambridge University Press, 2005).

Barnaby, Wendy, 'Australia: Environment down under', *Ambio* 12 (1983): 27–33.

Barnett, F. Oswald, '*The economics of the slums*'. Unpublished MComm thesis, University of Melbourne, 1931.

Barrett, Bernard, *The Inner Suburbs: The Evolution of an Industrial Area* (Melbourne: Melbourne University Press, 1971).

Barton, A. B., and Argue, J. R., 'A review of the application of water sensitive urban design (WSUD) to residential development in Australia', *Australasian Journal of Water Resources* 11 (2007): 31–40.

Barton, Ruth, 'Household technology in Western Australia 1900–1950', *Oral History Association of Australia Journal* 7 (1985): 108–29.

Bates, Daisy, *The Passing of the Aborigines: A Lifetime Spent Among the Natives of Australia* (New York: G.P. Puntam's Sons, 1939).

Battam, Harry, and Smith, Lindsay E., 'Report on review and analysis of albatross banding data held by the Australian bird and bat banding schemes', *Australian National Parks and Wildlife Service Research and Surveys Consultancy Agreement*, no. 138 (Sydney: Australian National Parks and Wildlife Service, 1994).

Bean, Charles E. W., *War Aims of a Plain Australian* (Sydney: Angus and Robertson, 1943).

Beder, Sharon, 'Getting into deep water: Sydney's extended ocean sewage outfalls', in Pam Scott (ed.), *A Herd of White Elephants: Australia's Science and Technology Policy* (Sydney: Hale and Iremonger, 1992), 62–74.

'Early environmentalists and the battle against sewers in Sydney', *Royal Australian Historical Society Journal* 76 (1990): 27–44.

Toxic Fish and Sewer Surfing: How Deceit and Collusion Are Destroying Our Great Beaches (Sydney: Allen and Unwin, 1989).

Beeliar Regional Park Final Management Plan (Perth: Department of Conservation and Land Management, 2006).

Belich, James, *Replenishing the Earth: The Settler Revolution and the Rise of the Anglo-World, 1783–1939* (Oxford: Oxford University Press, 2009).

Bell, Peter, *A History of the Queensland House* (Adelaide: Historical Research Pty Ltd, 2002).

Benac, David, 'The New Orleans lakefront: Nostalgia and the fate of New Urbanism', *Journal of Urban History* 41 (2015): 388–403.

Bertram, Nigel, and Murphy, Catherine, *In Time with Weather: Design Studies of 3 Australian Cities* (Perth: University of Western Australia Publishing, 2019).

Binnie International Ltd., *Development Study* (Perth: Metropolitan Water Supply and Sewerage and Drainage Board, 1977).

Birch, Tony, '"These children have been born in an abyss": Slum photography in a Melbourne suburb', *Australian Historical Studies* 35 (2004): 1–15.

Birnbauer, Bill, 'The big pong down under', *International Consortium of Investigative Journalists*, 13 February 2003, www.icij.org/investigations/waterbarons/big-pong-down-under/. Accessed 19 April 2021.

Blackburn, Kevin, 'The "consumer's ethic" of Australian advertising agencies, 1950–1965', *Journal of Australian Studies* 16 (1992): 60–74.

Blackbourn, David, *Conquest of Nature: Water, Landscape and the Making of Modern Germany* (New York: WW Norton and Company, 2006).

Blainey, Geoffrey, *The Tyranny of Distance: How Distance Shaped Australia's History* (Melbourne: Sun Books, 1966).

Bolton, Geoffrey, 'Perth: A foundling city', in Pamela Statham (ed.), *The Origins of Australia's Capital Cities* (Cambridge: Cambridge University Press, 1989), 141–55.

Bolton, Geoffrey, and Gregory, Jenny, *Claremont: A History* (Perth: University of Western Australia Press, 1999).

Bonyhady, Tim, *The Colonial Earth* (Melbourne: Melbourne University Press, 2002).

Places Worth Keeping: Conservationists, Politics, and Law (Sydney: Allen and Unwin, 1993).

Boot, H. M., 'Government and the colonial economies', *Australian Economic History Review* 38 (1998): 74–101.

Booth, Phillip, and Huxley, Margot, '1909 and all that: Reflections on the Housing, Town Planning, Etc. Act 1909', *Planning Perspectives* 27 (2012): 267–83.

Borland, Jeff, 'Microeconomic reform', in Simon Ville and Glenn Withers (eds.), *The Cambridge Economic History of Australia* (Melbourne: Cambridge University Press, 2015), 419–37.

Boughton, Walter C., 'Hydrology and water resources in Australia – the past', in *Proceedings of the 30th Hydrology and Water Resources Symposium* (Hobart: Conference Design, 2006), no pagination.

Boyce, James, *1835: The Founding of Melbourne & the Conquest of Australia* (Melbourne: Black Inc., 2011).

Boyd, Robin, *The Walls Around Us: The Story of Australian Architecture* (Melbourne: Children's Library Guild of Australia, 1962).

Australia's Home: Why Australians Built the Way They Did, 2nd ed. (Harmondsworth: Penguin, 1978).

Brady, D. K., and Cossins, Geoffrey, 'Factors affecting water consumption in Brisbane', *Institution of Engineers Australia, Queensland Division Technical Papers* 18 (1977): 1–6.

Bremner, A. J., and Chiffings, A. W., 'The Werribee Treatment Complex – an environmental perspective', *Water* 19 (1991): 22–24.

Brennan, Donna, Tapsuwan, Sorada, and Ingram, Gordon, 'The welfare costs of urban outdoor water restrictions', *Australian Journal of Agricultural and Resource Economics* 51 (2007): 243–61.

Broich, John, 'Engineering the Empire: British water supply systems and colonial societies, 1850–1900', *Journal of British Studies* 46 (2007): 346–65.

Broome, Richard, *Coburg: Between Two Creeks* (Melbourne: Lothian, 1987).

Aboriginal Victorians: A History since 1800 (Sydney: Allen & Unwin, 2005).

Aboriginal Australians: A History since 1788, 5th ed. (Sydney: Allen & Unwin, 2019).

Bropho, Robert, *Fringedweller* (Sydney: Alternative Publishing Co-operative, 1980).

Brunkova, Zdravka, and Shanahan, Martin, 'The Australian gold rushes 1850–1900: Elites, mineral ownership and democracy', in Andreas R. D. Sanders, Pål Thonstad Sandvik, and Espen Storli (eds.), *The Political Economy of Resource Regulation: An International and Comparative History, 1850–2015* (Vancouver: University of British Columbia Press, 2019), 23–44.

Bryant, Lyndall, *Investing in Australian Cities: The Legacy of the Better Cities Program* (Property Council of Australia, 2016)

Burgmann, Meredith, and Burgmann, Verity, *Green Bans, Red Union: Environmental Activism and the New South Wales Builders Labourers' Federation* (Sydney: UNSW Press, 1998).

Burroughs, Peter, 'Imperial institutions and the government of empire', in Andrew Porter (ed.), *The Oxford History of the British Empire*, Volume 3: *The Nineteenth Century* (Oxford: Oxford University Press, 1999), 170–97.

Bush, Judy, Miles, Barb, and Bainbridge, Brian, 'Merri Creek: Managing an urban waterway for people and nature', *Ecological Management and Restoration* 4 (2003): 170–79.

Butlin, N. G. (ed.), *Sydney's Environmental Amenity 1970–1975: A Study of the System of Waste Management and Pollution Control* (Canberra: Australian National University Press, 1976).

Our Original Aggression: Aboriginal Populations of Southeastern Australia, 1788–1850 (Sydney: Allen & Unwin, 1983).

Economics and the Dreamtime: A Hypothetical History (Melbourne: Cambridge University Press, 1993).

Butlin, N. G., Barnard, A., and Pincus, J. J., *Government and Capitalism: Public and Private Choice in Twentieth Century Australia* (Sydney: George Allen & Unwin, 1982).

Byrne, Denis, Goodall, Heather, and Cadzow, Allison, *Place-Making in National Parks: Ways that Australians of Arabic and Vietnamese Background Perceive and Use the Parklands Along the Georges River, NSW* (Sydney: NSW Office of Environment and Heritage, 2010).

Cadzow, Allison, Byrne, Denis, and Goodall, Heather, 'Waterborne: Vietnamese Australians and river environments in Vietnam and Sydney', *Transforming Cultures* 5 (2010): 112–42.

Cai, Wenju, Cowan, T, Rensch P. Van, and Weller, E., 'Did climate change-induced rainfall trends contribute to the Australian Millennium Drought?' *Journal of Climate* 27 (2014): 3145–68.

Cargeeg, G. C., Townley, L. R., Smith, G. R., Appleyard, S. J., and Smith, R. A., *Perth Urban Water Balance Study*, Vol. 1 (Perth: Water Authority of Western Australian, 1987).

Carroll, Brian, 'William Forgan Smith: Dictator or democrat?' in Denis Murphy, Roger Joyce, Margaret Cribb, and Rae Wear (eds.), *The Premiers of Queensland* (Brisbane: University of Queensland Press, 2003), 207–43.

Cathcart, Michael, *The Water Dreamers: The Remarkable History of Our Dry Continent* (Melbourne: Text, 2009).

Chandler, Tertius, *Four Thousand Years of Urban Growth: An Historical Census* (Lewistown, NY: St. David's University Press, 1987).

Chong, J., Herriman, J., Campbell, D., and White, S., *Review of Water Restrictions*, Vol. 1: *Review and Analysis* (Institute for Sustainable Futures, 2009).

Christensen, Joseph, 'Fremantle Doctor', in Jenny Gregory and Jan Gothard (eds.), *Historical Encyclopedia of Western Australia* (Perth: University of Western Australia Press, 2009).

'Industrial East Perth in growth and decline: The power station in its suburb', in Lenore Layman (ed.), *Powering Perth: A History of the East Perth Power Station* (Perth: Black Swan Press, 2011), 63–81.

City of Melbourne, *Urban Forest Strategy: Making a Great City Greener, 2012–2032* (Melbourne: City of Melbourne, 2011).

Clare, John, 'The Californian bungalow in Australia', *Historic Environment* 1 (1986): 19–39.

Clark, David, '"Worse than physic": Sydney's water supply 1788–1888', in Max Kelly (ed.), *Nineteenth-Century Sydney: Essays in Urban History* (Sydney: Sydney University Press, 1978), 54–65.

Clunies Ross, Ian, *Should We Plan for Peace* (Sydney: Consolidated, 1941).

Cole, John, *Shaping a City: Greater Brisbane 1925–1985* (Brisbane: William Brooks, 1984).

Collard, Leonard, 'Nyittiny: Cosmology of the Nuyngar of South-Western Australia', *Journal of Australian Indigenous Issues* 11 (2008): 3–11.

Commonwealth of Australia, *Census* (1911). www.abs.gov.au/AUSSTATS/abs@.nsf/DetailsPage/2112.01911.

(1921). www.abs.gov.au/AUSSTATS/abs@.nsf/DetailsPage/2111.01921.

Report of the Senate Select Committee on Water Pollution, Parliamentary Paper No. 98 (Canberra: Commonwealth Printing Office, 1970).

Commonwealth Housing Commission, *Final Report, 25th August 1944* (Sydney: Ministry of Post-War Reconstruction, 1944).
Connell, Daniel, *Water Politics of the Murray-Darling Basin* (Sydney: Federation Press, 2007).
Cook, Denise, *That Was My Home: Voices from the Noongar Camps in Fremantle and the Western Suburbs* (Perth: University of Western Australia Publishing, 2019).
Cook, Kevin, and Goodall, Heather, *Making Change Happen: Black and White Activists Talk to Kevin Cook about Aboriginal, Union and Liberation Politics* (Canberra: Australian National University Press, 2013).
Cook, Margaret, 'Vacating the floodplain: Urban property, engineering, and floods in Brisbane (1974–2011)', *Conservation and Society* 15 (2017): 344–54.
A River with a City Problem: A History of Brisbane Floods (Brisbane: University of Queensland Press, 2019).
'"Shaken but not stirred": The aftermath of urban disasters', in S. Mackinnon and M. Cook (eds.), *Disasters in Australia and New Zealand: Historical Approaches to Understanding Catastrophe* (Singapore: Palgrave Macmillan, 2020).
Cook, Margaret, and Spearritt, Peter, 'Water forever: Warragamba and Wivenhoe Dams', *Australian Historical Studies* 52, no. 2 (2021): 211–26.
Cooley, Heather, and Ajami, Newsha, 'Key issues for seawater desalination in California: Cost and financing', in Peter H. Gleick (ed.), *The World's Water: The Biennial Report on Freshwater Resources* (Washington, DC: Island Press, 2014).
Coutts, Andrew M., Tapper, Nigel J., Beringer, Jason, Loughnan, Margaret, and Demuzere, Matthias, 'Watering our cities: The capacity of Water Sensitive Urban Design to support urban cooling and improve human thermal comfort in the Australian context', *Progress in Physical Geography: Earth and Environment* 37 (2012): 2–28.
Cossins, G., 'One hundred years of Brisbane's water supply', *Institution of Engineers Australia: Brisbane Division Technical Papers* 7 (1966): 1–38.
Cox, Shaphan, Birdsall-Jones, Christina, Jones, Roy, Kerr, Thor, and Mickler, Steve, 'Indigenous persistence and entitlement: Noongar occupations in central Perth, 1988–1989 and 2012', *Journal of Historical Geography* 54 (2016): 13–23.
Coward, Dan, 'Reorganising waste management and pollution control, 1970–1975', in N. G. Butlin (ed.), *Sydney's Environmental Amenity 1970–1975: A Study of the system of Waste Management and Pollution Control* (Canberra: Australian National University Press, 1976).
Out of Sight: Sydney's Environmental History 1851–1981 (Canberra: Department of Economic History, ANU, 1988).
Crabb, Peter, 'Coping with change: New directions for the Sydney Water Board?' in Richard V. Cardew, John V. Langdale, and David C. Richard (eds.), *Why Cities Change: Urban Development and Economic Change in Sydney* (Sydney: Allen & Unwin 1982).

Crase, Lin, and Cooper, Bethany, 'The political economy of drought: Legacy and lessons from Australia's Millennium Drought', *Economic Papers* 36 (2017): 289–99.

Crosby, Alfred W., *Ecological Imperialism: The Biological Expansion of Europe 900–1900* (Cambridge: Cambridge University Press, 1986).

CSIRO and Bureau of Meteorology, *State of the Climate 2018* (Melbourne: CSIRO and Bureau of Meteorology, 2018).

Cullen, Peter, 'The Cinderella resource: Urban stormwater in a dry country', *Second International Symposium on Urban Stormwater Management, Melbourne, Australia*, 11–13 July 1995, 11–16.

Darian-Smith, Kate, *On the Home Front: Melbourne in Wartime: 1939–1945*, 2nd ed. (Melbourne: Melbourne University Press, 2009).

Daunton, M. J., *House and Home in the Victorian City: Working-Class Housing 1850–1914* (London: Edward Arnold, 1983).

David, Paul A., 'Path dependence, its critics and the quest for historical economics', in Pierre Garrouste and Stavros Ioannides (eds.), *Evolution and Path Dependence in Economic Ideas: Past and Present* (Cheltenham: Edward Elgar, 2001), 15–40.

Davison, Barbara, and Davison, Graeme, 'Suburban pioneers', in Graeme Davison, Tony Dingle, and Seamus O'Hanlon (eds.), *The Cream Brick Frontier: Histories of Australian Suburbia* (Melbourne: Monash Publications in Economic History No. 19, Monash University, 1995), 41–50.

Davison, Graeme, *The Rise and Fall of Marvellous Melbourne* (Melbourne: Melbourne University Press, 1978).

'The capital cities', in Graeme Davison, J. W. McCarty, and Ailsa McLeary (eds.), *Australians 1888* (Sydney: Fairfax, Syme & Weldon, 1987), 189–227.

'Colonial origins of the Australian home', in Patrick Troy (ed.), *A History of European Housing in Australia* (Cambridge: Cambridge University Press, 2000), 6–25.

'Gold-rush Melbourne', in Iain McCalman, Alexander Cook, and Andrew Reeves (eds.), *Gold: Forgotten Histories and Lost Objects of Australia* (Cambridge: Cambridge University Press, 2001), 52–66.

'Down the gurgler: Historical influences on Australian domestic water consumption', in Patrick Troy (ed.), *Troubled Waters: Confronting the Water Crisis in Australia's Cities* (Canberra: Australian National University Press, 2008), 37–65.

City Dreamers: The Urban Imagination in Australia (Sydney: NewSouth, 2016).

Davison, Graeme, and Dingle, Tony, 'Introduction: The view from the Ming Wing', in Graeme Davison, Tony Dingle, and Seamus O'Hanlon (eds.), *The Cream Brick Frontier: Histories of Australian Suburbia* (Melbourne: Monash Publications in Economic History No. 19, Monash University, 1995), 2–17.

Davison, Graeme and Lack, John, 'Planning the new social order: The Melbourne University Social Survey, 1941–43', *Australian and New Zealand Journal of Sociology* 16 (1981): 36–45.

Denholm, Tony, 'Adelaide: A Victorian bastide?' in Pamela Statham (ed.), *The Origins of Australia's Capital Cities* (Cambridge: Cambridge University Press, 1989), 179–96.

Dillon, P., Page, D., Vanderzalm, J., Pavelic, P., Toze, S., Bekele, E., Sidhu, J., Prommer, H., Higginson, S., Rinck-Pfeiffer, Regel, R., 'A critical evaluation of combined engineered and aquifer treatment systems in water recycling', *Water Science and Technology* 57 (2008): 753–62.

Dingle, Tony, 'Necessity the mother of invention, or do-it-yourself', in Patrick Troy (ed.), *A History of European Housing in Australia* (Melbourne: Cambridge University Press, 2000), 57–76.

Dingle, Tony, and Doyle, Helen, *Yan Yean: A History of Melbourne's Early Water Supply* (Melbourne: Public Record Office Victoria, 2003).

Dingle, Tony, and Rasmussen, Carolyn, *Vital Connections: Melbourne and Its Board of Works 1891–1991* (Melbourne: McPhee Gribble, 1991).

Director of Meteorology, *Brisbane Floods January 1974* (Canberra: Department of Science, Bureau of Meteorology, 1974).

Dobbie, Meredith, Morgan, Ruth, and Frost, Lionel, 'Overcoming abundance: Social capital and managing floods in inner Melbourne during the nineteenth century', *Journal of Urban History* 46 (2020): 33–49.

Dolnicar, Sara, and Hurlimann, Anna, 'Australians' water conservation behaviour and attitudes', *Australian Journal of Water Resources* 14 (2010): 43–53.

Donner, Erica et al., 'Wastewater monitoring for SARS-CoV-2', *Microbiology Australia* 42 (2021): 18–22.

Doucet, Michael J., and Weaver, John C., 'Material culture and the North American house: The era of the common man', *Journal of American History* 72 (1985): 560–87.

Dovers, Stephen, 'The rise and fall of the NSESD, or not?' *Australian Environmental Law News* 4 (1999): 30–41.

'Institutions for Sustainability'. *TELA: Environment, Economy and Society* 7, Australian Conservation Foundation, 2001: 1–33. https://openresearch-repository.anu.edu.au/bitstream/1885/40972/3/eeno101.pdf.

Dunstan, David, *Governing the Metropolis: Politics, Technology and Social Change in a Victorian City: Melbourne, 1850–1891* (Melbourne: Melbourne University Press, 1984).

Dovey, Kim, *Fluid City: Transforming Melbourne's Urban Waterfront* (Sydney: University of New South Wales Press, 2005).

Dowling, Robyn, and Power, Emma, 'Sizing home, doing family in Sydney, Australia', *Housing Studies* 27 (2012): 605–19.

Duncan, H. P., 'Water demand management in Melbourne', *Water Journal: Australian Water and Wastewater Association* 19 (1991): 26–28.

East Perth Redevelopment Authority, *Claisebrook Village: A Case Study in Urban Revitalisation* (Perth: East Perth Redevelopment Authority, 2011).

Ecologically Sustainable Development Steering Committee, *National Strategy for Ecologically Sustainable Development* (Canberra: Australian Government Printing Service, 1992).

Edmonds, Penelope, 'The intimate, urbanising frontier: Native camps and settler colonialism's violent array of spaces around early Melbourne', in Tracey Banivanua Mar and Penelope Edmonds, *Making Settler Colonial Space* (London: Palgrave Macmillan, 2010).

Edyvane, Karen, 'Issues in the South Australian marine environment', in Leon Zann and David Clement Sutton (eds.), *State of the Marine Environment Report for Australia: Technical Annex III State and Territory Issues* (Townsville: Great Barrier Reef Marine Park Authority, 1996), 61–88.

Engineering and Water Supply Department, *Water Resources Development 1836–1986* (Adelaide: Engineering and Water Supply Department, 1987).

Environmental Protection Authority (EPA), *Perth Metropolitan Desalination Proposal: Water Corporation of Australia* (Perth: Environmental Protection Authority, 2002).

Ergas, Henry, and Pincus, Jonathan, 'Infrastructure and colonial socialism', in Simon Ville and Glenn Withers (eds.), *The Cambridge Economic History of Australia* (Cambridge: Cambridge University Press, 2015), 222–44.

Evans, Raymond, *A History of Queensland* (Cambridge: Cambridge University Press, 2007).

Evans, Ray, and Saunders, Kay, 'No place like home: The evolution of the Australian housewife', in Kay Saunders and Ray Evans (eds.), *Gender Relations in Australia* (Sydney: Harcourt Brace Jovanovich, 1992), 176–90.

Ferrie, Joseph P., and Troesken, Werner, 'Water and Chicago's mortality transition, 1850–1925', *Explorations in Economic History* 45 (2008): 1–16.

Falkenmark, Malin, 'UN Water Conference: Agreement on goals and action plan', *Ambio* 6 (1977): 222–27.

Feikema, P. M., 'Influence of climate, fire severity and forest mortality on predictions of long term streamflow: Potential effect of the 2009 wildfire on Melbourne's water supply catchments', *Journal of Hydrology* 488 (2013): 1–16.

Field, C. A. R., and Sadler, Brian, *South West Regional Water Planning Study: Working Report* (Perth: Public Works Department, 1973).

Finlayson, C. Max, and Mitchell, David S., 'Management of salvinia (*Salvinia molesta*) in Australia', *Australian Weeds* 2, no. 2 (1982/83): 71–76.

Finlayson, B. L., McMahon, T. A., 'Australia v the world: A comparative analysis of streamflow characteristics', in Robin F. Warner (ed.), *Fluvial Geomorphology of Australia* (Sydney: Academic Press, 1988).

Fisher, Rod, 'From depredation to degradation: The Aboriginal experience at Moreton Bay 1842–1860', in Brisbane History Group, *The Aboriginal Presence 1824–1860* (Brisbane History Group Papers No. 11, 1992), 31–47.

Fishman, Robert, *Bourgeois Utopias: The Rise and Fall of Suburbia* (New York: Basic Books, 1987).

'Re-imagining Los Angeles', in Michael J. Dear, H. Erick Schockman, and Greg Hise (eds.), *Rethinking Los Angeles* (Thousand Oaks, CA: Sage, 1996), 251–61.

Fitzgerald, Shirley, *Sydney 1842–1992* (Sydney: Hale & Iremonger, 1992).

Fletcher, Tim D., Shuster, William, Hunt, William F., Ashley, Richard, Butler, David, Scott, Arthur, Towsdale, Sam, Barraud, Sylvie, Semadeni-Davies, Annette, Bertrand-Krajewski, Jean-Luc, Mikkelsen, Peter Steen, Rivard, Gilles, Uhl, Mathias, Dagenais, Danielle, Viklander, Maria, 'SUDS, LIDS, BMPs, WSUD and more – The evolution and application of terminology surrounding urban drainage', *Urban Water Journal* 12 (2015): 525–42.

Flinn, M. W., 'Introduction', in M. W. Flinn (ed.), *Report on the Sanitary Condition of the Labouring Population of Great Britain by Edwin Chadwick 1842* (Edinburgh: Edinburgh University Press, 1965).

Flood, J., *The Original Australians: Story of the Aboriginal People* (Sydney: Allen and Unwin, 2006).

Flynn, Joe, 'Designing a sustainable water industry', in C. B. Daniels and J. Hodgson (eds.), *Adelaide: Water of a City* (Adelaide: Wakefield Press, 2010), 489–510.

Foster, Robert, and Nettelback, Amanda, 'From protectorate to protection, 1836–1911', in Peggy Brock and Tom Gara (eds.), *Colonialism and Its Aftermath: A History of Aboriginal South Australia* (Adelaide: Wakefield Press, 2017), 27–40.

Fox, Charlie, 'Work organisation', in Patrick Bertola and Bobbie Oliver (eds.), *The Workshops: A History of the Midland Government Railway Workshops* (Perth: University of Western Australian Press, 2006), 81–105.

Fraser, Malcolm A. C., *Western Australian Year-Book for 1896–97* (Perth: Government Printer, 1898).

Freeman, Pamela, *Going Corporate: Restructuring the Sydney Water Board* (Sydney: Australian Graduate School of Management, University of New South Wales, 1998).

Freestone, Robert, 'An Imperial aspect: The Australasian town planning tour of 1914–15', *Australian Journal of Politics and History* 44 (1998): 159–76.

'Planning, housing, gardening: Home as a garden suburb', in Patrick Troy (ed.), *A History of European Housing in Australia* (Melbourne: Cambridge University Press, 2000), 125–41.

'Australian environmental planning origins and theories', in Jason Byrne et al. (eds.) *Australian Environmental Planning: Challenges and Future Prospects* (London: Routledge, 2014).

Urban Nation: Australia's Planning Heritage (Melbourne: CSIRO Publishing, 2010).

Freestone, Robert, and James, Peggy, 'Learning from LA: Australian responses to Los Angeles urbanism 1910–1960', *Journal of Planning History* 18 (2019): 44–65.

Frost, Lionel, *The New Urban Frontier: Urbanisation and City-Building in Australasia and the American West* (Sydney: New South Wales University Press, 1991).

'The contribution of the urban sector to Australian economic development before 1914', *Australian Economic History Review* 38 (1998): 42–73.

'Urbanisation', in Simon Ville and Glenn Withers (eds.), *The Cambridge Economic History of Australia* (Melbourne: Cambridge University Press, 2015), 245–63.

'Town and country in the Menzies years', in J. R. Nethercote (ed.), *Menzies: The Shaping of Modern Australia* (Brisbane: Connor Court, 2016), 231–55.

'Water technology and the urban environment: Water, sewerage, and disease in San Francisco and Melbourne before 1920', *Journal of Urban History* 46 (2020): 15–32.

Frost, Lionel, and Dingle, Tony, 'Sustaining suburbia: An historical perspective on Australia's growth', in Patrick Troy (ed.), *Australian Cities: Issues, Strategies and Policies for Urban Australia in the 1990s* (Cambridge: Cambridge University Press, 1995), 20–38.

Frost, Lionel, and O'Hanlon, Seamus, 'Urban history and the future of Australian cities', *Australian Economic History Review* 49 (2009): 1–18.

Frost, Lionel, and Shanahan, Martin, 'Domesticating water: How initial choices shaped water networks in three Australian cities', *Australian Historical Studies* 52, no. 2 (2021): 171–88.

Frost, Lionel, Gaynor, Andrea, Gregory, Jenny, Morgan, Ruth, O'Hanlon, Seamus, Spearritt, Peter, and Young, John, *Water, History and the Australian City: Urbanism, Suburbanism and Watering a Dry Continent, 1788–2015* (Melbourne: Cooperative Research Centre for Water Sensitive Cities, 2015), https://watersensitivecities.org.au/wp-content/uploads/2016/12/A2.1_2_2016_R5-19-12-2016-V2.pdf. Accessed 20 June 2021.

Frost, Warwick, *An Environmental History of Australian Rainforests until 1939* (Abingdon: Routledge, 2021).

Galishoff, Stuart, 'Triumph and failure: The American response to the urban water supply problem, 1860–1923', in Martin V. Melosi (eds.), *Pollution and Reform in American Cities, 1870–1930* (Austin, TX: University of Texas Press, 1980), 35–58.

Gandy, Matthew, *Concrete and Clay: Reworking Nature in New York City* (Cambridge, MA: MIT Press, 2002).

Garden, Don, '"Type 15", "Gengarry" and "Catalina": The changing space of the A.V. Jennings home in the 1960s', in Graeme Davison, Tony Dingle, and Seamus O'Hanlon (eds.), *The Cream Brick Frontier: Histories of Australian Suburbia* (Melbourne: Monash Publications in Economic History No. 19, Monash University, 1995), 140–53.

Gardiner, Anne, 'Do rainwater tanks herald a cultural change in household water use?' *Australasian Journal of Environmental Management* 17 (2010): 100–11.

Garnaut, Christine, *Colonel Light Gardens: Model Garden Suburb* (Sydney: Crossing Press, 2006).

Gautney, J. R., and Holliday, T. W., 'New estimations of habitable land area and human population size at the last glacial maximum', *Journal of Archaeological Science* 58 (2015): 103–12.

Gaynor, Andrea, *Harvest of the Suburbs: An Environmental History of Growing Food in Australian Cities* (Perth: University of Western Australia Press, 2006).

'Lawnscaping Perth: Water supply, gardens, and scarcity, 1890–1925', *Journal of Urban History* 46 (2020): 63–78.

Gaynor, Andrea, and Martin, Daniel Jan, 'Every fountain tells a story: Histories of civic water in Australia', *Australian Historical Studies* 52, no. 2 (2021): 189–210.

Gaynor, Andrea, Jennings, Philip, and Newman, Peter, 'A History of Roe 8 and the Perth Freight Link, 1955–2017', in Andrea Gaynor, Peter Newman, and

Philip Jennings (eds.), *Never Again: Reflections on Environmental Responsibility After Roe 8* (Perth: University of Western Australia Publishing, 2017).

Gibbons, Robert, 'The "fall of the giant": Trams versus trains and buses in Sydney, 1900–61', in Garry Wotherspoon (ed.), *Sydney's Transport: Studies in Urban History* (Sydney: Hale & Iremonger, 1983), 155–76.

Goodall, Heather, 'Riding the tide: Indigenous knowledge, history and water in changing Australia', *Environment and History* 14 (2008): 355–84.

'The river runs backwards', in Tim Bonyhady and Tom Griffiths (eds.), *Words for Country: Landscape and Language in Australia* (Sydney: University of New South Wales Press, 2002).

Waters of Belonging, Al-miyahu Tajma'unah: Arabic Australians and the Georges River Parklands (Sydney: University of Technology Sydney ePress, 2012).

Goodall, Heather, and Cadzow, Allison, 'Salt Pan Creek: Rivers as border zones within the colonial city', in Devleena Ghosh, Heather Goodall, and Hemelryk Donald (eds.), *Water, Sovereignty and Borders in Asia and Oceania* (New York: Routledge, 2008), 189–209.

'The people's national park: Working-class environmental campaigns on Sydney's Georges River, 1950–67', *Labour History* 99 (2010): 17–36.

Rivers and Resilience: Aboriginal People on Sydney's Georges River (Sydney: University of New South Wales Press, 2009).

Goodall, Heather, Cadzow, Allison, and Byrne, Denis, 'Mangroves, garbage, fishing: Bringing everyday ecology to Sydney's industrial Georges River', *Transforming Cultures eJournal* 5 (2010): 52–83.

Goodall, Heather, and Hekmat, Latifa, 'Talking to water: Memory, gender and environment for Hazara refugees in Australia', *International Review of Environmental History* 7 (2021): 81–99.

Goode, Brad, and Associates, *South West Regional Water Plan: Workshops Held within the Ngungar Community*, Reports for the Department of Water, Government of Western Australia, April 2008.

Gordon, Robert J., *The Rise and Fall of American Growth: The U.S. Standard of Living since the Civil War* (Princeton, NJ: Princeton University Press, 2016).

Grafton, R. Quentin, and Ward, Michael B., 'Prices versus rationing: Marshallian surplus and mandatory water restrictions', *Economic Record* 84 (2008): S57–S65.

Greenwood, Gordon, *Brisbane 1859–1959: A History of Local Government* (Brisbane: Brisbane City Council, 1959).

Gregory, Jenny, '*The manufacture of middle-class suburbia*'. Unpublished PhD thesis, University of Western Australia, 1989.

'Protecting middle-class suburbia: An ideal space for the citizens of interwar Perth', *Studies in WA History* 17 (1997): 77–91.

City of Light: A History of Perth since the 1950s (Perth: Perth City Council, 2003).

'Obliterating history? The transformation of inner city industrial suburbs', *Australian Historical Studies* 39 (2008): 91–106.

'"Let our watchword be 'order' and our beacon 'beauty'": Achieving town planning legislation in Western Australia', in Robert Freestone (ed.), *Cities, Citizens and Environmental Reform* (Sydney: Sydney University Press, 2009), 173–99.

'"A spirit of Bolshevism": Perth's water crisis of the 1920s', *Journal of Urban History* 46 (2020): 79–97.

Greif, Avner, and Mokyr, Joel, 'Cognitive rules, institutions, and economic growth: Douglass North and beyond', *Journal of Institutional Economics* 13 (2017): 25–52.

Greig, Alistair, *The Stuff Dreams Are Made of: Housing Provision in Australia 1945–1960* (Melbourne: Melbourne University Press, 1995).

Grigg, Trevor J., *A Comprehensive Evaluation of the Proposed Wivenhoe Dam on the Brisbane River* (Brisbane: Co-ordinator General's Department, 1977).

Groth, Paul, 'Workers'-cottage and minimal-bungalow districts in Oakland and Berkeley, California, 1870–1945', *Urban Morphology* 8 (2004): 13–25.

Gunn, A. W., Werbeloff, L., Chesterfield, C., Hammer, K., and Rogers, B. C., *Benchmarking, Envisioning and Transition Planning for a Water Sensitive Adelaide: Final Case Report* (Melbourne, Australia: Cooperative Research Centre for Water Sensitive Cities, 2017).

Haebich, Anna, *For Their Own Good: Aborigines and Government in the Southwest of Western Australia, 1900–1940* (Perth: University of Western Australia Press, 1992).

Hall, Tony, *The Life and Death of the Australian Backyard* (Melbourne: CSIRO Publications, 2010).

Hamer, David, *New Towns in the New World: Images and Perceptions of the Nineteenth-Century Urban Frontier* (New York: Columbia University Press, 1990).

Hamlin, Christopher, *Cholera: The Biography* (New York: Oxford University Press, 2009).

Hammerton, Marianne, *Water South Australia: A History of the Engineering and Water Supply Department* (Adelaide: Wakefield Press, 1986).

Hancke, Steve H., and Davis, Robert K, 'Demand management through responsive pricing', *Journal of the American Water Works Association* 63 (1971): 555–60.

Hanebuth, Till, and Stattegger, Karl, 'Rapid flooding of the Sunda Shelf: A late glacial sea-level record', *Science* 288, no. 5468 (2000): 1033–35.

Hardin, Garrett, 'The tragedy of the commons', *Science* 162 (1968): 1243–48.

Hartley, Ken, *Independent Audit of the Bolivar Wastewater Treatment Plant to Determine the Causes of a Major Odour Event* (Brisbane: Uniquest, 1997).

Hartley, Richard G., *River of Steel: A History of the Western Australian Goldfields and Agricultural Water Supply 1895–2003* (Perth: Access Press, 2007).

Harvey, David, *A Brief History of Neoliberalism* (Oxford: Oxford University Press, 2005).

Hawkins, Gay, 'Shit in public', *Australian Humanities Review (Online)* (2004): 31–32, http://australianhumanitiesreview.org/2004/04/01/shit-in-public/?utm_source=rss&utm_medium=rss&utm_campaign=shit-in-public. Accessed 19 April 2021.

'Waste in Sydney: Unwelcome returns', *PMLA* 22 (2007): 348–51.

Head, Brian W., 'Managing urban water crises: Adaptive policy responses to drought and flood in Southeast Queensland, Australia', *Ecology and Society* 19 (2014), 33.

Head, L. M., Farbotko, C., Gibson, C., Gill, N., and Waitt, Gordon R., 'Zones of friction, zones of traction: The connected household in climate change and sustainability policy', *Australian Journal of Environmental Management* 20 (2013): 351–62.

Head, Lesley, 'Nature, networks and desire: Changing cultures of water in Australia', in Patrick Troy (ed.), *Troubled Waters: Confronting the Water Crisis in Australia's Cities* (Canberra: ANU Press, 2008), 67–80.

Hector, Donald, 'Sydney's water sewerage and drainage system', *Journal and Proceedings of the Royal Society of New South Wales* 144 (2011): 3–25.

Hill, M. R., *Housing Finance in Australia 1945–1956* (Melbourne: Melbourne University Press: 1959).

Hillyer, Vivienne, 'Bennett House: Aboriginal Heritage as real estate in East Perth', *Urban Policy and Research* 19 (2001): 147–82.

Hilmer, Fred G., *National Competition Policy Review* (Canberra: Australian Government Printer, 1993).

Hirst, John, *The Strange Birth of Colonial Democracy: New South Wales 1848–1884* (Sydney: Allen & Unwin, 1988).

Holdsworth, Deryck, 'Regional distinctiveness in an industrial age: Some Californian influences on British Columbia housing', *Review of Canadian Studies* 12 (1982): 64–81.

Holland, Graham, 'The comfortable house', in Patrick Troy (ed.), *A History of European Housing in Australia* (Melbourne: Cambridge University Press, 2000), 197–217.

Emoh Ruo: Owner Building in Sydney (Sydney: Hale & Iremonger, 1988).

Holmes, Katie, Martin, Susan K., and Mirmohamadi, Kylie, *Reading the Garden: The Settlement of Australia* (Melbourne: Melbourne University Press, 2008).

Honert, Robin C. van den, and McAneney, John, 'The 2011 Brisbane floods: Causes, impacts and implications', *Water* 3 (2011): 1149–73.

Hoskins, Ian, 'Constructing time and space in the garden suburb', in Sarah Ferber, Chris Healy, and Chris McAuliffe (eds.), *Beasts of Suburbia: Reinterpreting Cultures in Australian Suburbs* (Melbourne: Melbourne University Press, 1994), 1–17.

Hull, W., *Report of the Select Committee of the Legislative Council on the Aborigines, 1858–9 (Victoria)* (Melbourne: Committee of the Legislative Council on the Aborigines. Government of Australia, 1859).

Hulme, Mike, *Why We Disagree about Climate Change: Understanding Controversy, Inaction and Opportunity* (Cambridge: Cambridge University Press, 2009).

Hunt, H.E., *'Address', in Seminar H2O: Domestic Extravagance* (Perth: University of Western Australia Press, 1975).

Perth's Early Water Supplies (Perth: Western Australia Division, Institute of Engineers Australia, 1984).

Hunt, Su-Jane, *Water: The Abiding Challenge* (Perth: Metropolitan Water Board, 1980).

Hunt, Su-Jane, and Bolton, Geoffrey, 'Cleansing the dunghill: Water supply and sanitation in Perth 1878–1912', *Studies in Western Australian History* 2 (1978): 1–17.

Hunt, Suzanne, 'Lawn mower', in Richard Aitken and Michael Looker (eds.), *The Oxford Companion to Australian Gardens* (Melbourne: Oxford University Press, 2002).

Hunter, Boyd H., and Carmody, David, 'Estimating the Aboriginal population of early colonial Australia: The role of chickenpox reconsidered', *Australian Economic History Review* 55 (2015): 112–38.

Hurlimann, Anna, and Sara Dolnicar, 'When public opposition defeats alternative water projects – The case of Toowoomba, Australia', *Water Research* 44 (2010): 287–97.

Hutton, Drew, and Connors, Libby, *A History of the Australian Environment Movement* (Cambridge: Cambridge University Press, 1999).

Industry Commission, *Water Resources and Wastewater Disposal*, Inquiry Report No. 26 (Canberra: Australian Government Publishing Service, 1992).

Irish, Paul, *Hidden in Plain View: The Aboriginal People of Coastal Sydney* (Sydney: NewSouth, 2017).

Irish, Paul, and Ingrey, Michael, 'Marking their footsteps', in Christine Hansen and Kathleen Butler (eds.), *Exploring Urban Identities and History* (Canberra; Australian Institute of Aboriginal and Torres Strait Islander Studies, 2013), 75–90.

Jackson, P. et al., *A Water Plan for a Growing Melbourne: Water Supply Strategy Review* (Melbourne: Melbourne Metropolitan Water Board, 1986).

Jackson, Sue, and Moggridge, Bradley, 'Indigenous water management', *Australasian Journal of Environmental Management* 26 (2019): 193–96.

Jacobs, Jane M., *Edge of Empire: Postcolonialism and the City* (London: Routledge, 1996).

Jacobson, Charles David, *Ties that Bind: Economic and Political Dilemmas of Urban Utility Networks, 1800–1990* (Pittsburgh, PA: University of Pittsburgh Press, 2000).

Jewitt, Sarah, 'Geographies of shit: Spatial and temporal variations in attitudes towards human waste', *Progress in Human Geography* 35 (2011): 608–26.

Johnston, Ross, and Gregory, Helen, 'Brisbane: Making it work', in Pamela Statham (ed.), *The Origins of Australia's Capital Cities* (Cambridge: Cambridge University Press, 1989), 252–68.

Jones, David S., 'Kaurna cultural heritage', in C. B. Daniels and J. Hodgson (eds.), *Adelaide: Water of a City* (Adelaide: Wakefield Press, 2010), 131–54.

Jones, Jonathan E., Earles, T. Andrew, Fassman, Elizabeth A., Herricks, Edwin E., Ubonas, Ben, and Clary, Jane K., 'Urban storm-water regulations – Are impervious area limits a good idea?' *Journal of Environmental Engineering* 131 (2005), 176–79.

Jones, Rebecca, *Slow Catastrophes: Living with Drought in Australia* (Clayton: Monash University Publishing, 2017).

Joseph, M., 'Pesticides', *Medical Journal of Australia* 1 (1964): 49–50.

Joy, C. S., and Hickson, W. J., 'Of sewers and drains', in N. G. Butlin (ed.), *Sydney's Environmental Amenity 1970–1975: A Study of the System of Waste Management and Pollution Control* (Canberra: Australian National University Press, 1976).

Joyce, Patrick, *The Rule of Freedom: Liberalism and the Modern City* (London: Verso, 2003).

Kaika, Maria, and Swyngedouw, Erik, 'Fetishizing the modern city: The phantasmagoria of urban technological networks', *International Journal of Urban and Regional Research* 24 (2000): 120–38.

'The urbanization of nature: Great promises, impasse, and new beginnings', in Gary Bridge and Sophie Watson (eds.), *The New Blackwell Companion to the City* (Chichester, West Sussex: Wiley-Blackwell, 2013), 96–107.

Kaika, Maria, 'Interrogating the geographies of the familiar: Domesticating nature and constructing the autonomy of the modern home', *International Journal of Urban and Regional Research* 28 (2004): 265–86.

'Constructing scarcity and sensationalising water politics: 170 days that shook Athens', *Antipode* 35 (2003): 919–54.

City of Flows: Modernity, Nature and the City (London: Routledge, 2005).

Karskens, Grace, *Inside the Rocks: The Archaeology of a Neighbourhood* (Sydney: Hale & Iremonger, 1999).

The Colony: A History of Early Sydney (Sydney: Allen & Unwin, 2009).

'Floods and flood-mindedness in early colonial Australia', *Environmental History* 21 (2016): 315–42.

'Water dreams, earthen histories: Exploring urban environmental history at the Penrith Lakes Scheme and Castlereagh, Sydney', *Environment and History* 13 (2007): 115–54.

Karskens, Grace, Wilkins, Erin, Seymour, Jasmine, Watson, Leanne, and Wright, Rhiannon, "Friday essay: How a long-lost list is helping us remap Darug place names and culture on Dyarubbin, the Hawkesbury River', *The Conversation*, 18 December 2020.

Kellogg Brown, and Root Pty Ltd, *Metropolitan Adelaide Stormwater Management: Part B: Stormwater Harvesting and Use* (Adelaide: Metropolitan Adelaide Stormwater Management Steering Committee, 2004).

Kelly, Ian, '*The development of housing in Perth (1890–1915)*'. Unpublished MArch thesis, University of Western Australia, 1992.

Kelly, Max, *Paddock Full of Houses: Paddington 1840–1890* (Sydney: Doak Press, 1978).

'Picturesque and pestilential: The Sydney slum observed 1860–1900', in Max Kelly (ed.), *Nineteenth-Century Sydney: Essays in Urban History* (Sydney: Sydney University Press, 1978), 66–80.

'Pleasure and profit: The eastern suburbs come of age 1919–1929', in Jill Roe (ed.), *Twentieth Century Sydney: Studies in Urban and Social History* (Sydney: Hale & Iremonger, 1980), 1–17.

Kerkhove, Ray, *Aboriginal Campsites of Greater Brisbane* (Brisbane Boolarong Press, 2015).

'Aboriginal camps as urban foundations? Evidence from southern Queensland', *Aboriginal History* 42 (2018): 147–72, https://press-files.anu.edu.au/downloads/press/n4634/html/article07.xhtml. Accessed 19 April 2021.

Khan, Stuart, *Drinking Water through Recycling: The Benefits and Costs of Supplying Direct to the Distribution System* (Melbourne: Australian Academy of Technological Sciences and Engineering, 2013).

Kilner, David, *'The evolution of South Australian urban housing policy'*. PhD thesis, University of Adelaide, 1988.

King, Anthony D., *The Bungalow: The Production of a Global Culture* (London: Routledge and Kegan Paul, 1984).

Kingsford, Richard, Fairweather, Peter G., Geddes, Michael C., Lester, Rebecca E., Sammut, Jesmond, and Walker, Keith F., 'Engineering a crisis in a Ramsar wetland: The Coorong, Lower Lakes and Murray Mouth, Australia', *Marine and Freshwater Research* 62 (2011): 255–65.

Kingston, Beverly, 'Mackellar, Isobel Marion Dorothea (1885–1968)', *Australian Dictionary of Biography*, National Centre of Biography, Australian National University, https://adb.anu.edu.au/biography/mackellar-isobel-marion-dorothea-7383/text12835, published first in hardcopy 1986. Accessed 16 April 2021.

Krestschmer, Peter, *Managed Aquifer Recharge Schemes in the Adelaide Metropolitan Area*, DEWNR Technical report 2017/22 (Adelaide: Department of Environment, Water and Natural Resources, 2017).

Kwaymullina, Ambelin, 'Research, ethics and Indigenous people: An Australian Indigenous perspective on three threshold considerations for respectful engagement', *AlterNative: An International Journal of Indigenous Peoples*, 12 (2016): 437–449.

et al., 'Living texts: A perspective on published sources, Indigenous research methodologies and Indigenous worldviews', *International Journal of Critical Indigenous Studies* 6, no. 1 (2013), 1–13, https://ijcis.qut.edu.au/article/view/106.

Lack, John, *A History of Footscray* (Melbourne: Hargreen, 1991).

Lake, Marilyn, 'Historical homes', *Australian Historical Studies* 24 (1991): 46–54.

La Nauze, Robert D., *Engineer to Marvellous Melbourne: The Life and Times of William Thwaites* (Melbourne: Australian Scholarly Publishing, 2011).

Langford, K. J., and Heeps, D. P., *A Water Supply Strategy for Melbourne: Planning for an Uncertain Future* (Melbourne: Melbourne Metropolitan Water Board, 1982).

'A demand strategy for Victoria', *National Engineering Conference, Melbourne, 2–8 March 1985* (Canberra: Institution of Engineers, Australia, 1985).

Langton, Marcia, 'Earth, wind, fire and water: The social and spiritual construction of water in Aboriginal societies', in Bruno David et al. (eds.), *The Social Archaeology of Australian Indigenous Studies* (Canberra: Aboriginal Studies Press, 2008).

Laverty, John R., *The Making of a Metropolis: Brisbane 1823–1925* (Brisbane: Boolarong Press, 2009).

Lawson, Ronald, *Brisbane in the 1890s: A Study of an Australian Urban Society* (Brisbane: University of Queensland Press, 1973).
Leigh, Carol, 'Bassendean: A rich and varied heritage', *Early Days: Journal of the Royal Western Australian Historical Society* 13 (2009): 339–53.
Lemon, Andrew, *The Northcote Side of the River* (Melbourne: Hargreen, 1983).
Leroux, Anke D., Martin, Vance L., and Zheng, Hao, 'Addressing water shortages by force of habit', *Resource and Energy Economics* 53 (2018): 42–61.
Lewi, Hannah, 'Whose heritage: The contested site of the Swan Brewery', *Fabrications* 15 (2005): 43–61.
Lewis, Milton J., *The People's Health: Public Health in Australia, 1950 to the Present*, Vol. 2 (Westport, CT: Praeger, 2003).
Lewis, Sophie C., King, Andrew D., and Mitchell, Daniel, 'Australia's unprecedented future temperature extremes under Paris limits to warming', *Geophysical Research Letters* 44 (2017): 9947–56.
Lindsay, J., and Supski, S., 'Changing household water consumption practices after drought in three Australian cities', *Geoforum* 84 (2017), 51–58.
Lloyd, Clem, Troy, Patrick, and Shreiner, Shelley, *For the Public Health: The Hunter District Water Board, 1892–1992* (Melbourne: Longman Cheshire, 1992).
Lloyd, C. J., *Either Drought or Plenty: Water Development and Management in New South Wales* (New South Wales Department of Water Resources, 1988).
Lloyd, Clem, 'Water, sewerage and drainage', in Patrick Troy (ed.), *Technological Change and the City* (Sydney: Federation Press, 1995), 54–77.
Lloyd, Sara D., *Water Sensitive Urban Design in the Australian Context, Synthesis of a Conference Held 30–31 August 2000, Melbourne Australia, Technical Report 1/7* (Melbourne: CRC for Catchment Hydrology, 2001).
Lockwood, Christine, 'Early encounters on the Adelaide Plains and Encounter Bay', in Peggy Brock and Tom Gara (eds.), *Colonialism and Its Aftermath: A History of Aboriginal South Australia* (Adelaide: Wakefield Press, 2017), 65–81.
Low, Tim, *The New Nature* (Melbourne: Penguin Random House, 2017).
Lynch, Cassie *'Split', in Stories of Perth* (Sydney: Seizure Books, 2018).
Lynch, Spence, and Pearson, *Parameters for the River: A Prospectus for Rehabilitation of Parramatta River, Commissioned by the National Trust of Australia (NSW), in Conjunction with the Parramatta River Conservation Co-ordination Committee* (Sydney: National Trust of Australia NSW, 1976).
Macintyre, Dobson, and Associates 2003, *Report on an Ethnographic, Ethnohistorical, Archaeological and Underwood Avenue Bushland Project Area, Shenton Park*. A report prepared for University of Western Australia, 2003.
Macintyre, Stuart, *Australia's Boldest Experiment: War and Reconstruction in the 1940s* (Sydney: New South, 2015).
'Owners and tenants: The Commonwealth Housing Commission and post-war housing, 1943–1949', *Australian Economic History Review* 58 (2018): 265–82.
McCalman, Janet, *Struggletown: Public and Private Life in Richmond 1900–1965* (Melbourne: Melbourne University Press, 1984).

McCarty, J. W., 'Australian capital cities in the nineteenth century', *Australian Economic Review* 10 (1970): 107–37.

McClellan, P., *Sydney Water Inquiry Final Report* (Sydney: Sydney Water Inquiry Secretariat, 1998).

McCloskey, Deidre N., *Bourgeois Equality: How Ideas, Not Capital or Institutions, Enriched the World* (Chicago: University of Chicago Press, 2016).

McDonald, Edward, Coldrick, Bryn, and Christensen, William, 'The green frog and desalination: A Nyungar metaphor for the (mis-)management of water resources, Swan Coastal Plain, Western Australia', *Oceania Publications* 78 (2008), 62–75.

McFarlane, Don, 'Will Perth have enough water for its diverse needs in a drying climate?', in Sharon Biermann, Doina Olaru, and Valerià Paül (eds.), *Planning Boomtown and Beyond* (Perth: University of Western Australia Publishing 2016), 209–38.

Mackellar, Dorothea, *My Country and Other Poems* (Melbourne: Viking O'Neill, 1993).

McLean, Ian W., *Why Australia Prospered: The Shifting Sources of Economic Growth* (Princeton, NJ: Princeton University Press, 2013).

McNeill, J. R., *Mosquito Empires: Ecology and War in the Greater Caribbean, 1620–1914* (Cambridge: Cambridge University Press, 2010).

McPhaden, Michael J., Santoso, Agus, and Cai, Wenju, 'Introduction: El Nino Southern Oscillation in a changing climate', in M. J. McPhaden, A. Santoso, and W. Cai (eds.), *El Nino Southern Oscillation in a Changing Climate* (Hoboken: American Geophysical Union, 2021), 1–19.

Marsden, John, and Pickering, Phil, *Securing Australia's Urban Water Supplies* (Melbourne: Marsden Jacob Associates, 2006).

Martin, Daniel, and London, Geoffrey, 'The deep waters of Perth', in Nigel Bertram and Catherine Murphy (eds.), *In Time with Water: Design Studies of 3 Australian Cities* (Perth: University of Western Australia Publishing, 2019).

Massola, A., *Bunjil's Cave: Myths, Legends and Superstitions of the Aborigines of South-East Australia* (Melbourne: Lansdowne Press, 1968).

Mauger, Graeme, *Planning Future Sources for Perth's Water Supply* (Perth: Water Authority of Western Australia, 1987).

May, Cathie, *Changes They've Seen: The City and People of Bayswater 1827–1997* (Perth: City of Bayswater, 1997).

Mayne, Alan, 'A just war: The language of slum representation in twentieth-century Australia', *Journal of Urban History* 22 (1995): 75–107.

Meagher, Sara J., 'The food resources of the Aborigines of the south-west of Western Australia', *Records of the Western Australian Museum* 3 (1974): 14–65.

Meehan, Katherine, Ormerod, Kerri Jean, and Moore, Sarah A., 'Remaking waste as water? The governance of recycled effluent for potable water supply', *Water Alternatives* 6 (2013): 67–85.

Mein Smith, Phillipa, and Frost, Lionel, 'Suburbia and infant death in late nineteenth- and early twentieth-century Adelaide', *Urban History* 21 (1994): 251–72.

Meinig, D. W., *The Shaping of America: A Geographical Perspective on 500 Years of History: Volume 3 Transcontinental America, 1850–1913* (New Haven, CT: Yale University Press, 1998).

Melbourne and Metropolitan Board of Works, *Report on Environmental Study into Thomson Dam and Associated Works*, Vol. 2: *Environmental Implications and Conclusions* (Melbourne: Advocate Press, 1975).

Melbourne Water, *Port Phillip Bay Environmental Study: The Findings, 1992–1996* (Melbourne: Melbourne Water, 1996).

Melosi, Martin V., *The Sanitary City: Urban Infrastructure in America from Colonial Times to the Present* (Pittsburgh: University of Pittsburgh Press, 2000).

Precious Commodity: Providing Water for America's Cities (Pittsburgh: University of Pittsburgh Press, 2011).

Menck, Clare, *A Thematic History of Government Housing in Western Australia: Final Report* (Perth: Western Australia Department of Housing, 2014).

Menzies, Robert, 'The forgotten people (1942)', in Judith Brett (ed.), *Robert Menzies' Forgotten People* (Sydney: Pan Macmillan, 1992), 5–14.

Merrett, D. T., 'Economic growth and well-being: Melbourne 1870–1914: A comment', *Economic Record* 53 (1977): 262–71.

'Australian capital cities in the twentieth century', in J. W. McCarty and C. B. Schedvin (eds.), *Australian Capital Cities: Historical Essays* (Sydney: University of Sydney Press, 1978), 171–98.

Merrett, David, and Ville, Simon, 'Tariffs, subsidies, and profits: A re-assessment of structural change in Australia 1901–39', *Australian Economic History Review* 31 (2011): 46–70.

Metropolitan Water Authority, *Domestic Water Use in Perth, Western Australia* (Perth: Metropolitan Water Centre, 1985).

Middelmann, Miriam, Rodgers, Simon, White, Justin, Cornish, Lisa, and Zoppou, Christopher, 'Riverine flood hazard', in Trevor Jones, Miriam Middelmann, and Neil Corby (eds.), *Natural Hazard Risk in Perth, Western Australia: Comprehensive Report* (Canberra: Geoscience Australia, 2005), 102–06.

Miles, Keith R., *Geology and Underground Water Resources of the Adelaide Plains Area, Bulletin* 27 (Adelaide: Geological Survey of South Australia, 1952).

Millward, Robert, 'The political economy of urban utilities', in Martin Daunton (ed.), *The Cambridge Urban History of Britain*, Vol. 3: *1840–1950* (Cambridge: Cambridge University Press, 2000), 315–49.

Ministry of Conservation, *Assessment of the Thomson Dam and Associated Works* (Melbourne: Ministry of Conservation, 1975).

Mitchell, Timothy, *Rule of Experts: Egypt, Techno-Politics, Modernity* (Berkeley: University of California, 2002).

Moggridge, Bradley J., and Thompson, Ross M., 'Cultural value of water and western water management: An Australian Indigenous perspective'. *Australian Journal of Water Resources* 25 (2021), 4–14.

Moore, G. F., *Descriptive Vocabulary of the Language in Common Use amongst the Aborigines of Western Australia* (London: Walbrook, 1884).

Mokyr, Joel, *The Enlightened Economy: An Economic History of Britain 1700–1850* (New Haven, CT: Yale University Press, 2009).

Mokyr, Joel, and Stein, Rebecca, 'Science, health, and household technology: The effect of the Pasteur revolution on consumer demand', in Timothy F. Bresnahan and Robert J. Gordon (eds.), *The Economics of New Goods* (Chicago: University of Chicago Press, 1997), 183–84.

Moore, Sarah A., 'Garbage matters: Concepts in new geographies of waste', *Progress in Human Geography* 36 (2012): 780–99.

Moore, Tony, 'Life is not meant to be easy', in James Walter and Tony Moore (eds.), *What Were They Thinking? The Politics of Ideas in Australia* (Sydney: University of New South Wales Press, 2010).

Morgan, Ruth, 'A thirsty city: An environmental history of water supply and demand in 1970s Perth, Western Australia', *Studies in Western Australian History* 27 (2011): 81–97.

Running Out? Water in Western Australia (Perth: University of Western Australia Press, 2015).

Moy, Candice, 'Rainwater tank households: Water savers or water users?' *Geographical Research* 50 (2012).

Muller, Edward K., and Tarr, Joel A., 'The interaction of natural and built environments in the Pittsburgh landscape', in Joel A. Tarr (ed.), *Devastation and Renewal: An Environmental History of Pittsburgh and Its Region* (Pittsburgh, University of Pittsburgh Press, 2003), 11–40.

Mulligan, Martin, and Hill, Stuart B., *Ecological Pioneers: A Social History of Australian Thought and Action* (Cambridge: Cambridge University Press, 2002).

Murphy, John, and Probert, Belinda, '"Anything for the house": Recollections of post-war suburban dreaming', *Australian Historical Studies* 26 (2004): 275–93.

Murray-Wallace, C. V., 'Pleistocene coastal stratigraphy, sea-level highstands and neotectonism of the southern Australian passive continental margin – a review', *Journal of Quaternary Science* 17 (2002): 469–89, https://doi.org/10.1002/jqs.717.

Musgrave, W., 'The political economy of resource use: Water', in J. A. Sinden (ed)., *The Natural Resources of Australia: Prospects and Problems for Development* (Sydney: Angus and Robertson, 1974).

Myers, Roland, 'Swimming pools', *The Bulletin*, 28 October 1972: 63–64, https://nla.gov.au/nla.obj-1335022273/view?sectionId=nla.obj-1501011433&partId=nla.obj-1335095152#page/n62/mode/1up. Accessed 19 April 2021.

Nannup, Noel, 'Walk with the Waugal' [videorecording], https://youtu.be/eAnPK-jBxeY, 31 December 2019.

Nash, Linda, *Inescapable Ecologies: A History of Environment, Disease, and Knowledge* (Berkeley: University of California Press, 2003).

Neutze, Max, *Urban Development in Australia: A Descriptive Analysis* (Sydney: George Allen & Unwin, 1981).

Funding Urban Services: Options for Physical Infrastructure (Sydney: Allen & Unwin, 1997).

'The provision of hydraulic services', in Patrick N. Troy (ed.), *Serving the City: The Crisis in Australia's Urban Services* (Sydney: Pluto Press, 1999), 60–79.

New South Wales Premier's Department, *Sydney Water Inquiry Final Report*, 1998.

NHMRC, *Water Fluoridation and Human Health in Australia: Questions and Answers* (Canberra: NHMRC, 2017).

Australian Drinking Water Guidelines Paper 6, National Water Quality Management Strategy (Canberra: NHMRC and National Resource Management Ministerial Council, 2011).

North, Douglass C., *Transaction Costs, Institutions, and Economic Performance* (San Francisco: International Center for Economic Growth, 1992).

Nunn, Patrick, and Reid, Nicholas, 'Aboriginal memories of inundation of the Australian coast dating from more than 7000 years ago', *Australian Geographer* 47 (2016): 11–47.

O'Connell, B. P., 'Major city water supplies', *Water Journal* 4, no. 4 (1977): 14–15.

O'Connor, Rory, Bodney, Corrie, and Little, L., *Preliminary Report on the Survey of Aboriginal Areas of Significance in the Perth Metropolitan and Murray River Regions* (Perth, Heritage Council of WA, 1985).

O'Connor, R. et al., *Report on an Investigation into Aboriginal Significance of Wetlands and Rivers in the Perth-Bunbury Region* (Perth: Western Australian Water Resources Council, 1989).

Petrie, Constance, *Tom Petrie's Reminiscences of Early Queensland* (Brisbane: Watson Ferguson and Co., 1904).

Pitrans, Harry, 'Stormwater management in Adelaide: A new partnership'. *Seventh National Local Government Engineering Conference: Effective Use of Technology* (1993). https://search.informit.org/doi/10.3316/informit.381822142446921.

O'Gorman, Emily, *Flood Country: An Environmental History of the Murray-Darling Basin* (Melbourne: CSIRO Publishing, 2012).

O'Gorman, Emily, and Gaynor, Andrea, 'More-than-human histories', *Environmental History* 25 (2020): 711–35.

O'Gorman, Emily, and Morgan, Ruth A., 'Fluid terrains: Approaches in environmental history, *Australian Historical Studies* 52, no. 2 (2021): 141–170.

O'Hanlon, Seamus, and Spearritt, Peter, 'From water engineers to financial engineering: Water provision in Australia's east coast capital cities, 1945–2015', *Journal of Urban History* 46 (2020): 98–112.

O'Hanlon, Seamus, *City Life: The New Urban Australia* (Sydney: ReadHowYouWant, 2020).

O'Rourke, Kevin H., and Williamson, Jeffrey G., 'Once more: When did globalization begin?' *European Review of Economic History* 8 (2004): 109–17.

Ostrom, Elinor, 'A behavioral approach to the rational choice theory of collective action', *American Political Science Review* 92 (1998): 1–22.

'Collective action and the evolution of social norms', *Journal of Economic Perspectives* 14 (2000): 137–58.

'The challenge of common-pool resources', *Environment: Science and Policy for Sustainable Development* 50 (2008): 8–21.

Panza, Laura, and Williamson, Jeffrey G., 'Australian squatters, convicts, and capitalists: Dividing up a fast-growing pie, 1821–71', *Economic History Review* 72 (2019): 568–94.

Parliament of South Australia, *Report of the Select Committee on SA Water* (Adelaide: Legislative Council Parliamentary Papers, 2008–9).

Paterson, John, 'The privatisation issue: Water utilities', in P. W. Abelson (ed.), *Privatisation: An Australian Perspective* (Sydney: Australian Professional Publications, 1987).

Peck, Adam, Barrett, Geoff, and Williams, Matt, *The 2019 Great Cocky Count: A Community-Based Survey for Carnaby's Black-Cockatoo (Calyptorhynchus latirostris), Baudin's Black-Cockatoo (Calyptorhynchus baundinii) and Forest Red-Tailed Black-Cockatoo (Calyptorhynchus banksia naso)* (Perth: BirdLife Australia, 2019).

Peel, Mark, 'Between the houses: Neighbouring and privacy', in P. Troy (ed.), *A History of European Housing in Australia* (Melbourne: Cambridge University Press, 2000), 269–86.

Penrose, Helen, *Werribee Farm: A History* (Melbourne: Melbourne Water Corporation, 2001).

Pigram, John J., *Issues in the Management of Australia's Water Resources* (Melbourne, Longman Cheshire, 1986).

'Outdoor recreation and access to countryside: Focus on the Australian experience', *Natural Resources Journal* 21 (1981): 107–24.

Pikusa, Stefan, *The Adelaide House 1836–1901: The Evolution of Principal Dwelling Types* (Adelaide: Wakefield Press, 1986).

Prakash, Gyan, *Another Reason: Science and the Imagination in Modern India* (Princeton, NJ: Princeton University Press, 1999).

Pollett, Chris, 'Public water supply schemes and groundwater response', in Whelan, Brian (ed.), *The Groundwater Resources of the Swan Coastal Plain*, Proceedings of a Symposium Held at the University of Western Australia, Perth, WA, 21 and 22 May 1981.

Potter, Emily, 'The gathering storm: Adelaide's olive trees in a changing climate', *Griffith Review* 55 (2017): 225–32.

Powell, Joseph M., *Watering the Western Third: Water, Land, and Community in Western Australia, 1826–1998* (Perth: Water and Rivers Commission, 1998).

Queensland Government, *Somerset Dam* (Brisbane: Queensland State Public Relations Bureau, 195?).

Quiggin, John, 'Contracting out: Promise and performance', *Economic and Labour Relations Review* 13 (2002): 88–104.

Race, Kane, '"Frequent sipping": Bottled water, the will to health and the subject of hydration', *Body and Society* 18(2012): 72–98.

Radbone, I., and Hamnett, S., 'Making the most of Adelaide's water', in C. A. Brebbia, J. F. Martin-Duque, and L. C. Wadhwa (eds.), *The Sustainable City II* (Southampton: WIT Press, 2002).

Radcliffe, John C., Page, Declan, Naumann, Bruce, and Dillon, Peter, 'Fifty years of water sensitive urban design, Salisbury, South Australia', *Frontiers Environmental Science Engineering* 11 (2017): 1–10.

Radcliffe, John, and Page, Declan, 'Water reuse and recycling in Australia – history, current situation and future perspectives', *Water Cycle* 1 (2020): 19–40.

Radcliffe, John, 'History of water sensitive urban design/low impact development adoption in Australia and internationally', in Ashok K. Sharma, Ted Gardner, and Don Begbie (eds.), *Approaches to Water Sensitive Urban Design: Potential, Design, Ecological Health, Urban Greening, Economics, Policies, and Community Perceptions* (Amsterdam: Elsevier 2019), 1–24.

Randolph, Bill, 'The changing Australian city: New patterns, new policies and new research needs', *Urban Policy and Research* 22 (2004): 481–93.

Rawson, Michael, *Eden on the Charles: The Making of Boston* (Cambridge, MA: Harvard University Press, 2010).

Rechner, Judy, 'The Queensland Workers Dwelling 1910–1940', *Royal Historical Society of Queensland Journal* 15 (1994): 266–71.

'*Houses for Queenslanders of small means: Workers' dwellings in old Coorparoo Shire 1910–40*'. Unpublished MA thesis, University of Queensland, 1998.

Reid, D. J., and Bone, E. K., 'The rise and fall of oyster cultivation in the highly urbanised Georges River estuary, Sydney, Australia: A review of lessons learned', *Regional Studies in Marine Science* 35 (2020): 101246.

Reiger, Kerreen, *The Disenchantment of the Home* (Melbourne: Oxford University Press, 1985).

'All but the kitchen sink: On the significance of domestic science and the silence of social theory', *Theory and Society* 16 (1987): 497–526.

Ridley, Matt, *The Rational Optimist: How Prosperity Evolves* (London: Fourth Estate, 2010).

Robertson, Robert W., and Keady, P., *Holroyd Municipal Swimming Centres Management and Financial Review: A Project of the 1984–85 Local Government Development Program* (Canberra: Australian Government Public Service, 1987).

Rolls, J. M., 'Stormwater management in metropolitan Adelaide: Technical innovation and institutional rigidity', *Second International Symposium on Urban Stormwater Management, Melbourne, Australia*, 11–13 July 1995, 87–91.

Ronalds, Albert F., *Future Water Supply of the Melbourne Metropolitan Area* (Melbourne: Melbourne Metropolitan Board of Works, 1962).

Rosenberg, Charles, *The Cholera Years: The United States in 1832, 1849, and 1866* (Chicago: University of Chicago Press, 1987).

Sainty, G., McCorkelle, G., and Julien, M., 'Control and spread of alligator weed *Alternanthera philoxeroides* (Mart.) Griseb., in Australia: Lessons for other regions', *Wetlands Ecology and Management* 5, no. 3 (1997): 195–201.

Satur, Paul A., '*Social inequality and water sensitive cities in Australia*'. Unpublished PhD thesis, Monash University, 2017.

Saw, William A., 'Some aspects of town planning', *Journal and Proceedings of the Royal Society of Western Australia* 5 (1918–19), 39–61.

Schedvin, C. B., *Australia and the Great Depression: A Study of Economic Development and Policy in the 1920s and 1930s* (Sydney: Sydney University Press, 1970).

Schlunke, A., Lewis J., and Fane, S., 'Analysis of Australian opportunities for more efficient toilets', prepared for the Australian Government Department of the Environment, Water, Heritage and the Arts, Institute for Sustainable Futures (Sydney: University of Technology, February 2008).

Schmidt, Jeremy, *Water: Abundance, Scarcity and Security in the Age of Humanity* (New York: New York University Press, 2017).

Scott, James C., *Seeing Like a State: How Certain Schemes to Improve the Human Condition Have Failed* (New Haven, CT: Yale University Press, 1998).

Scott, Pam (ed.), *A Herd of White Elephants: Australia's Science and Technology Policy* (Sydney: Hale and Iremonger, 1992).

Searle, Glen, 'In search of the elusive triple bottom line: Turbulent infrastructure policy at the Sydney Water Board', *Proceedings of the State of the Australian Cities National Conference 2007* (Adelaide: University of South Australia, 2007), 99–109.

Searle, G., 'The long-term urban impacts of the Sydney Olympic Games', *Australian Planner* 49 (2012): 195–202.

Seddon, George, *Swan River Landscapes* (Perth: University of Western Australia Press, 1970).

'The Australian back yard', in Ian Craven (ed.), *Australian Popular Culture* (Cambridge: Cambridge University Press, 1994), 22–35.

Swan Song: Reflections on Perth and Western Australia (Perth: University of Western Australia Press, 1995).

Landprints (Melbourne: Cambridge University Press, 1997).

Senate Environment and Communications References Committee, *Stormwater Management in Australia* (Canberra: Senate Environment and Communications References Committee, 2015).

Shanahan, Martin P., 'Wealth and welfare', in Simon Ville and Glenn Withers (eds.), *The Cambridge Economic History of Australia* (Melbourne: Cambridge University Press, 2015), 489–510.

Sharma, Ashok K., Gardner, Ted, and Begbie, Don (eds.), *Approaches to Water Sensitive Urban Design: Potential, Design, Ecological Health, Urban Greening, Economics, Policies, and Community Perceptions* (Duxford: Elsevier, 2019).

Sharp, Liz, 'Water demand management in England and Wales: Constructions of the domestic water user', *Journal of Environmental Planning and Management* 49 (2006): 869–89.

Shaw, Kate (ed.), *Planning Practice 1998: The Best and Worst Examples of City Planning and Development* (North Melbourne: People's Committee for Melbourne, 1998).

Sheil, Christopher, '"An incomplete hypothesis": Deregulation of water and sewerage in Australia', *Utilities Policy* 12 (2004): 153–64.

Water's Fall: Running the Risks with Economic Rationalism (Sydney: Pluto, 2000).

Shove, Elizabeth, *Comfort, Cleanliness and Convenience in the Social Organization of Normality* (Oxford: Berg, 2004).

Simms, Andrew, Hamilton, Stacey, and Biswas, Wahidul K. Biswas, 'Carbon footprint assessment of Western Australian Groundwater Recycling Scheme', *Environmental Management* 59 (2017): 557–70.

Sinclair, W. A., 'Capital formation', in C. Forster (ed.), *Australian Economic Development in the Twentieth Century* (London: George Allen & Unwin, 1970), 11–65.

Sinden, John A. (ed)., *The Natural Resources of Australia: Prospects and Problems for Development* (Sydney: Angus and Robertson, 1974).

Sinnett, Frederick, *An Account of the Colony of South Australia: Prepared for Distribution at the International Exhibition of 1862* (London: Robert K. Burt, 1862).

Siolda, James, 'Razing San Francisco: The 1906 disaster as a natural experiment in urban redevelopment', *Journal of Urban Economics* 89 (2015): 48–61.

Smith, Anita et al., 'Indigenous knowledge and resource management as World Heritage values: Budj Bim cultural landscape, Australia', *Archaeologies* 15 (2019): 285–313.

Smith, David I., *Water in Australia: Resources and Management* (Melbourne: Oxford University Press, 1998).

Smith, F. B., *The People's Health 1830–1910* (Canberra: Australian National University Press, 1979).

Smith, Jim, *The Aboriginal People of the Burragorang Valley*, 2nd ed. (Sydney: Blue Mountains Education and Research Trust, 2018).

Smith, K., 'Trends in water resource management', *Progress in Physical Geography* 3 (1979): 236–54.

Smith, Len, McCalman, Janet, Anderson, Ian, Smith, Sandra, Evans, Joanne, McCarthy, Gavan, and Beer, Jane, 'Fractional identities: The political arithmetic of Aboriginal Victorians', *Journal of Interdisciplinary History* 38 (2008): 533–51.

Sofoulis, Zoë, 'Big water, everyday water: A sociotechnical perspective', *Continuum: Journal of Media and Cultural Studies* 19 (2005): 445–63.

Soll, David, *Empire of Water: An Environmental and Political History of the New York City Water Supply* (Ithaca, NY: Cornell University Press, 2013).

South Australia Engineering and Water Supply Department, *Water South Australia: Managing the Resource into the Next Century* (Adelaide: The Department, 1989).

Southcott, R. V., Crosby, N. D., and Stenhouse, N.S., 'Studies on the epidemiology of the 1947–1948 epidemic of poliomyelitis in South Australia', *Medical Journal of Australia* 2 (1949), 481–95.

Spearritt, Peter, 'Suburban cathedrals: The rise of the drive in shopping centre', in Graeme Davison, Tony Dingle and Seamus O'Hanlon (eds.), *The Cream Brick Frontier: Histories of Australian Suburbia* (Melbourne: Monash Publications in Economic History No. 19, Monash University, 1995), 88–107.

Sydney's Century: A History (Sydney: UNSW Press, 1999).

'The water crisis in Southeast Queensland: How desalination turned the region into carbon emission heaven', in Patrick Troy (ed.), *Troubled Waters: Confronting Water Crisis in Australia's Cities* (Canberra: ANU E Press, 2008), 19–36.

'The 200 km City: Brisbane, the Gold Coast, and Sunshine Coast', *Australian Economic History Review* 49 (2009): 87–106.

Spirn, Anne, 'New Urbanism and the environment', *Places* 13 (2000): 44.

Spoehr, John D., Quiggin, John, Wilson, Lou, and Purse, Kevin, *State of Secrecy: Outsourcing: Promise and Performance* (Adelaide: Centre for Labour Research, *Adelaide University*, 2002).

Stanes, Elyse R., Klocker Natascha, and Gibson, Christopher R., 'Young adult households and domestic sustainabilities', *Geoforum* 65 (2015): 46–58.

Stanford, J. D. et al., 'Sea-level probability for the last deglaciation: A statistical analysis of far-field records', *Global and Planetary Change* 79 (2010): 193–203.

Stannage, C. T., *The People of Perth: A Social History of Western Australia's Capital City* (Perth: Perth City Council, 1979).

Starr, Kevin, *Inventing the Dream: California through the Progressive Era* (New York: Oxford University Press, 1985).

Steele, John, *Aboriginal Pathways in South East Queensland and the Richmond River* (Brisbane: University of Queensland Press, 1983).

Summerville, Jennifer A., Adkins, Barbara, and Kendall, Gavin, 'Community participation, rights and responsibilities: The governmentality of sustainable development policy in Australia', *Environment and Planning C: Government and Policy* 26 (2008): 696–711.

Sydney Water, Sydney Water's submission for the mid-term review of its Operating Licence, 2002, www.ipart.nsw.gov.au/files/sharedassets/website/trimholdingbay/sydney_water_corporation_-_s4964.pdf.

Syme, Geoff J., Kantola, S. J., and Thomas, J. F., 'Water resources and the quarter-acre block', in R. Thorne and S. J. Arden (eds.), *People and the Man-Made Environment* (Sydney, University of Sydney, 1980).

Syme, G. J., 'Summary of the education/persuasion mechanism', in *Proceedings of the National Workshop on Urban Water Demand Management* (Perth: WA Water Resources Council, 1986).

Syme, G. L., and Kantola, S. J., 'Investment in private bores: Underground water usage from a household perspective', in B. R. Whelan (ed.), *Groundwater Resources of the Swan Coastal Plain* (Perth: CSIRO, 1981), 453–76.

Tarr, Joel A., 'The separate vs. combined sewer problem: A case study in urban technology design choice', *Journal of Urban History* 5 (1979): 308–39.

The Search for the Ultimate Sink: Urban Pollution in Historical Perspective (Akron, OH: University of Akron Press, 1996).

Taylor, Elizabeth, 'The "desiccator difficulty": Surprise, indignation and the local politics of planning for sanitary technology in nineteenth century Melbourne', *Planning Perspectives* 35 (2020): 433–55.

Taylor, Katherine Selena, Moggridge, Bradley J., and Poelina, Anne, 'Australian Indigenous Water Policy and the impacts of the ever-changing political cycle', *Australasian Journal of Water Resources* 20 (2016): 132–47.

Taylor, Vanessa, and Trentmann, Frank, 'Liquid politics: Water and politics of everyday life in the modern city', *Past and Present* 211 (2011): 199–241.

Teichelmann, C. G., and Schürmann, C. W., *Outlines of a Grammar, Vocabulary, and Phraseology of the Aboriginal Language of South Australia, Spoken by the Natives in and for Some Distance Around Adelaide* (Adelaide: Teichelmann and C. W. Schürmann, 1840), http://nla.gov.au/nla.obj-1160965034. Accessed 9 March 2021.

World Meteorological Organization, *The Dublin Statement and Report of the Conference: International Conference on Water and the Environment:*

Development Issues for the 21st Century, 26–31 January, Dublin, Ireland, Principle 4.
Thomas, Merredith, 'East Perth 1884–1904: A suburban society', in J. W. McCarty and C. B. Schedvin (eds.), *Australian Capital Cities: Historical Essays* (Sydney: Sydney University Press, 1978), 144–58.
Thompson, Susannah, *Beyond Matta Garup: A History of Victoria Park* (Town of Victoria Park in association with Centre for WA History, University of Western Australia, 2012).
Thorne, Ross, and Arden, Stuart J. (eds.), *People and the Man-Made Environment* (Sydney: University of Sydney, 1980).
Tomes, Nancy, 'The private side of public health: Sanitary science, domestic hygiene, and the germ theory', *Bulletin of the History of Medicine* 64 (1990): 509–39.
Trelease, Frank J., 'Policies for water law: Property rights, economic forces and public regulation', *Natural Resources Journal* 5 (1865): 1–48.
Troy, Patrick (ed.), *Troubled Waters: Confronting the Water Crisis in Australia's Cities* (Canberra: Australian National University Press, 2008).
Troy, Patrick, Holloway, Darren, and Randolph, Bill, *Water Use and the Built Environment: Patterns of Water Consumption in Sydney* (Sydney: UNSW City Futures Research Centre, 2005).
Troy, Patrick, 'The life and times of the Chadwickian solution', in Patrick Troy (ed.), *Troubled Waters Confronting Water Crisis in Australia's Cities* (Canberra: ANU E Press, 2008).
Accommodating Australians: Commonwealth Government Involvement in Housing (Sydney: Federation Press, 2012).
Turner, A., White, S., Kazaglis, A., and Simard, S., 'Have we achieved the savings? The importance of evaluations when implementing demand management', *Water Science and Technology: Water Supply* 7 (2007): 203–10.
Tyrrell, Ian, *River Dreams: The People and the Landscape of the Cooks River* (Sydney: University of Sydney Press, 2018).
United States Bureau of the Census, *Mortality Statistics 1910* (Washington, DC: Government Printing Office, 1913).
Valtin, Heinz, and Gorman, Sheila A., '"Drink at least eight glasses of water a day": Really? Is there scientific evidence for "8x8"?', *American Journal of Physiology: Regulatory, Integrative and Comparative Physiology* 283 (2002): R993–R1004.
Vamplew, Wray (ed.), *Australians: Historical Statistics* (Sydney: Fairfax, Syme & Weldon, 1987).
Veenstra, Gerry, 'Location, location, location: Contextual and compositional health effects of social capital in British Columbia, Canada', *Social Science & Medicine* 60 (2005): 2059–71.
Victoria, *Select Committee on the Sewerage of and Supply of Water for Melbourne, Sewerage and Supply: Report of the Select Committee on the Sewerage of and Supply of Water for Melbourne* (Melbourne: J. Ferres, Government Printer, 1853).
Viggers, James, Weaver, Haylee, and Lindenmayer, David, *Melbourne's Water Catchments: Perspectives on a World-Class Water Supply* (Melbourne: CSIRO Publishing, 2013).

Viska, John, *A Guide to Conserving and Interpreting Gardens in Western Australia* (Perth: Australian Garden History Society [WA Branch], 2007).

Walling, Edna, *A Gardener's Log, First Published 1948, Republished 1985*, ed. Margaret Barrett (Melbourne: Anne O'Donovan, 1985).

Wallis, Philip, Birrell, Bob, Griggs, Dave, Healey, Ernest, Langford, John, and Stanley, Janet, *Melbourne's Water Situation: The Opportunity for Diverse Solutions* (Melbourne: Monash Sustainability Institute, Monash University, 2009).

Walder, Edwin, 'Water supply and sewerage in an expanding metropolis', *Australian Journal of Public Administration* 28 (1968): 171–80. https://doi/abs/10.1111/j.1467-8500.1969.tb00738.x. Accessed 19 April 2021.

Walters, B. et al., 'Different engineering solutions to environmental stress on two rivers in the Sydney Basin', *National Environmental Engineering Conference, Melbourne, 17–19 March 1986* (1986), 52–59.

Walton, Andrea, and Hume, Margee, 'Creating positive habits in water conservation: The case of the Queensland Water Commission and the Target 140 campaign', *International Journal of Nonprofit and Voluntary Sector Marketing* 16 (2011): 215–24.

Wapshott, Nicholas, *Keynes/Hayek: The Clash That Defined Modern Economics* (Melbourne: Scribe, 2011).

Water and Rivers Commission, *Environmental Management of Groundwater Abstraction from the Gnangara Mound 2000–2003* (East Perth: Water and Rivers Commission, 2004).

Water Corporation, 'Renewable energy generator – Beenyup Wastewater Treatment Plant', *Water Corporation Ongoing Works* (2020), www.watercorporation.com.au/Outages-and-works/Ongoing-Works/Renewable-Energy-Generator—Beenyup-Wastewater-Treatment-Plant. Accessed 19 April 2021.

Ward, Anthony, '*The development of Melbourne in the interwar years*'. Unpublished PhD thesis, Monash University, 1984.

Webber, Kimberley, '*Romancing the machine: The enchantment of domestic technology in the Australian home, 1850–1914*'. Unpublished PhD thesis, University of Sydney, 1997.

Wentworth Group of Concerned Scientists, 'Australia's Climate is Changing Australia', November 2006, https://wentworthgroup.org/wp-content/uploads/2013/10/Australia%e2%80%99s-Climate-is-Changing-Australia.pdf. Accessed 19 April 2021.

Western Australian Water Resources Council, *Proceedings of the National Workshop on Urban Water Demand Management* (Perth: Western Australian Water Resources Council, 1987).

Whelan, Brian (ed.), 'The groundwater resources of the Swan coastal plain', Proceedings of a Symposium Held at the University of Western Australia, Perth, WA, 21 and 22 May, 1981.

Whish-Wilson, David, Dravnieks, Kim, Bartlett, Alison, and Gaynor, Andrea, 'A twenty-first century grassroots protest', in Andrea Gaynor, Peter Newman, and Philip Jennings (eds.), *Never Again: Reflections on*

Environmental Responsibility After Roe 8 (Perth: University of Western Australia Publishing, 2018).

Whitington, Don, *The Effluent Society: Pollution in Australia* (Melbourne: Thomas Nelson, 1970).

Whittington, Vera, *Gold and Typhoid: Two Fevers: A Social History of Western Australia 1891–1900* (Perth: University of Western Australia Press, 1988).

Wilenski, Peter, *Directions for Change: An Interim Report, Parliamentary Paper, No. 290* (Sydney: NSW Parliament, 1978).

Williamson, D. R., 'Statistics of water use', *Proceedings of Seminar H_2O: Domestic Extravagance* (Perth: Water Research Foundation of Australia, 1975).

Williamson, Jeffrey G., *Coping with City Growth during the British Industrial Revolution* (Cambridge: Cambridge University Press, 1990).

Trade and Poverty: When the Third World Fell Behind (Cambridge, MA: MIT Press, 2011).

Williams, Michael, *The Making of the South Australian Landscape: A Study in the Historical Geography of Australia* (London: Academic Press, 1974).

Withers, Glenn, 'Immigration and economic fluctuations: An application to nineteenth-century Australia', *Australian Economic History Review* 17 (1977): 131–49.

Wohl, Anthony S., *Endangered Lives: Public Health in Victorian Britain* (London: J.M. Dent & Sons, 1983).

Wood, Lavinia E. E., *Brisbane City Council Swimming Pools: The Next Twenty Years* (Brisbane: Brisbane City Council, 1992).

Woodruff, Philip, *Two Million South Australians* (Adelaide: Peacock Publications, 1984).

Wooltorton, Sandra et al., 'Living water: Groundwater and wetlands in Gnangara, Noongar boodjar', *PAN: Philosophy, Activism, Nature* 14 (2018): 5–23.

World Meteorological Organization, *International Conference on Water and the Environment, Development Issues for the 21st Century, 26–31 January 1992: The Dublin Statement and the Report of the Conference* (Geneva: WMO, 1992).

Wotherspoon, Garry, 'The "Sydney interest" and the rail, 1860–1900', in Max Kelly (ed.), *Nineteenth-Century Sydney: Essays in Urban History* (Sydney: Sydney University Press, 1978), 12–23.

Zelko, Frank, 'Optimizing nature: Invoking the "natural" in the struggle over water fluoridation', *History of Science* 57 (2019): 518–39.

Index

Studies in Environment and History

Other Books in the Series *(continued)*:

John L. Brooke, *Climate Change and the Course of Global History: A Rough Journey*
Paul Josephson, et al., *An Environmental History of Russia*
Emmanuel Kreike, *Environmental Infrastructure in African History: Examining the Myth of Natural Resource Management*
Gregory T. Cushman, *Guano and the Opening of the Pacific World: A Global Ecological History*
Sam White, *The Climate of Rebellion in the Early Modern Ottoman Empire*
Edmund Russell, *Evolutionary History: Uniting History and Biology to Understand Life on Earth*
Alan Mikhail, *Nature and Empire in Ottoman Egypt: An Environmental History*
Richard W. Judd, *The Untilled Garden: Natural History and the Spirit of Conservation in America, 1740–1840*
James L. A. Webb, Jr., *Humanity's Burden: A Global History of Malaria*
Myrna I. Santiago, *The Ecology of Oil: Environment, Labor, and the Mexican Revolution, 1900–1938*
Frank Uekoetter, *The Green and the Brown: A History of Conservation in Nazi Germany*
Matthew D. Evenden, *Fish versus Power: An Environmental History of the Fraser River*
Alfred W. Crosby, *Ecological Imperialism: The Biological Expansion of Europe, 900–1900, second edition*
Nancy J. Jacobs, *Environment, Power, and Injustice: A South African History*
Edmund Russell, *War and Nature: Fighting Humans and Insects with Chemicals from World War I to Silent Spring*
Adam Rome, *The Bulldozer in the Countryside: Suburban Sprawl and the Rise of American Environmentalism*
Judith Shapiro, *Mao's War against Nature: Politics and the Environment in Revolutionary China*
Andrew Isenberg, *The Destruction of the Bison: An Environmental History*
Thomas Dunlap, *Nature and the English Diaspora*
Robert B. Marks, *Tigers, Rice, Silk, and Silt: Environment and Economy in Late Imperial South China*
Mark Elvin and Tsui'jung Liu, *Sediments of Time: Environment and Society in Chinese History*
Richard H. Grove, *Green Imperialism: Colonial Expansion, Tropical Island Edens and the Origins of Environmentalism, 1600–1860*
Thorkild Kjærgaard, *The Danish Revolution, 1500–1800: An Ecohistorical Interpretation*
Donald Worster, *Nature's Economy: A History of Ecological Ideas, second edition*
Elinor G. K. Melville, *A Plague of Sheep: Environmental Consequences of the Conquest of Mexico*
J. R. McNeill, *The Mountains of the Mediterranean World: An Environmental History*
Theodore Steinberg, *Nature Incorporated: Industrialization and the Waters of New England*
Timothy Silver, *A New Face on the Countryside: Indians, Colonists, and Slaves in the South Atlantic Forests, 1500–1800*

Michael Williams, *Americans and Their Forests: A Historical Geography*
Donald Worster, *The Ends of the Earth: Perspectives on Modern Environmental History*
Robert Harms, *Games against Nature: An Eco-Cultural History of the Nunu of Equatorial Africa*
Warren Dean, *Brazil and the Struggle for Rubber: A Study in Environmental History*
Samuel P. Hays, *Beauty, Health, and Permanence: Environmental Politics in the United States, 1955–1985*
Arthur F. McEvoy, *The Fisherman's Problem: Ecology and Law in the California Fisheries, 1850–1980*
Kenneth F. Kiple, *The Caribbean Slave: A Biological History*

CPSIA information can be obtained
at www.ICGtesting.com
Printed in the USA
BVHW041803110522
636809BV00001B/10

9 781108 831581